Springer Theses

Recognizing Outstanding Ph.D. Research

Aims and Scope

The series "Springer Theses" brings together a selection of the very best Ph.D. theses from around the world and across the physical sciences. Nominated and endorsed by two recognized specialists, each published volume has been selected for its scientific excellence and the high impact of its contents for the pertinent field of research. For greater accessibility to non-specialists, the published versions include an extended introduction, as well as a foreword by the student's supervisor explaining the special relevance of the work for the field. As a whole, the series will provide a valuable resource both for newcomers to the research fields described, and for other scientists seeking detailed background information on special questions. Finally, it provides an accredited documentation of the valuable contributions made by today's younger generation of scientists.

Theses are accepted into the series by invited nomination only and must fulfill all of the following criteria

- They must be written in good English.
- The topic should fall within the confines of Chemistry, Physics, Earth Sciences, Engineering and related interdisciplinary fields such as Materials, Nanoscience, Chemical Engineering, Complex Systems and Biophysics.
- The work reported in the thesis must represent a significant scientific advance.
- If the thesis includes previously published material, permission to reproduce this must be gained from the respective copyright holder.
- They must have been examined and passed during the 12 months prior to nomination.
- Each thesis should include a foreword by the supervisor outlining the significance of its content.
- The theses should have a clearly defined structure including an introduction accessible to scientists not expert in that particular field.

More information about this series at http://www.springer.com/series/8790

Akinori Tanaka

Superconformal Index on $RP^2 \times S^1$ and 3D Mirror Symmetry

Doctoral Thesis accepted by
Osaka University, Osaka, Japan

Springer

Author
Dr. Akinori Tanaka
Particle Physics Theory Group, Physics Department
Osaka University
Suita, Osaka
Japan

Supervisor
Prof. Yutaka Hosotani
Particle Physics Theory Group, Physics Department
Osaka University
Suita, Osaka
Japan

ISSN 2190-5053 ISSN 2190-5061 (electronic)
Springer Theses
ISBN 978-981-10-9346-3 ISBN 978-981-10-1398-0 (eBook)
DOI 10.1007/978-981-10-1398-0

Printed on acid-free paper

This Springer imprint is published by Springer Nature
The registered company is Springer Science+Business Media Singapore Pte Ltd.

Supervisor's Foreword

Quantum field theory (QFT) is a universal language to describe modern physics. Its validity spans vast branches, such as particle physics, condensed matter physics, statistical physics, mathematical physics, and string theory. In particle physics $SU(3)$ color gauge theory, quantum chromodynamics (QCD), describes strong interactions among quarks and gluons, whereas $SU(2) \times U(1)$ unified gauge theory of Glashow–Salam–Weinberg describes electroweak interactions of quarks and leptons. These theories have been solidly confirmed by many experiments.

Although QFT is very powerful, we are far from fully understanding it particularly in the strong coupling regime. In many situations we rely on perturbation analysis, but this misses many interesting and rich non-perturbative dynamics of QFT. In the path-integral formulation of QFT it is quite difficult to conduct an exact evaluation. How to conduct the path-integral exactly has been a long-standing problem in QFT.

Recently there has been a significant development in solving this problem in QFT with supersymmetry (SUSY). With a large enough amount of SUSY, a new technique called the localization method can be developed with which one can conduct the path-integral exactly by the analytical method. This opens up many new directions. For example, by performing the path-integral exactly and evaluating the partition functions of the two theories, one can establish the duality between these two QFTs: nontrivial equivalence between (apparently) different theories.

Dr. Akinori Tanaka is an expert on this subject. Throughout his Ph.D. work, he has mastered this localization technique and revealed many nontrivial aspects of QFT from the exact results he obtained using the localization technique, and this thesis is one such revelation. In this thesis, Akinori evaluated the superconformal index on $RP^2 \times S^1$ using the localization method and directly showed the duality between the XYZ-model and supersymmetric quantum electrodynamics (SQED). The thesis contains a transparent introduction of the subject with both physics and mathematics backgrounds and will serve as a good introductory monograph for researchers.

It is my honor to introduce Akinori's work for publication in the Springer Theses series. His work was nominated as an Outstanding Physics Ph.D. Thesis of the Fiscal Year 2014 by the Department of Physics, Graduate School of Science, Osaka University.

Osaka, Japan
March 2016

Prof. Yutaka Hosotani

Parts of this thesis have been published in the following journal articles:

- H. Mori and A. Tanaka, "Varieties of Abelian mirror symmetry on $RP^2 \times S^1$," *J. High Energy Phys.* **02** (2016) 088, arXiv:1512.02835 [hep-th].
- H. Mori, T. Morita, and A. Tanaka, "Single-flavor Abelian mirror symmetry on $RP^2 \times S^1$," in 9th International Symposium on Quantum theory and symmetries (QTS-9) Yerevan, Armenia, July 13-18, 2015. arXiv:1510.08598 [hep-th].
- A. Tanaka, H. Mori, and T. Morita, "Abelian 3d mirror symmetry on $RP^2 \times S^1$ with $N_f = 1$," *J. High Energy Phys.* **09** (2015) 154, arXiv:1505.07539 [hep-th].
- A. Tanaka, H. Mori, and T. Morita, "Superconformal index on $RP^2 \times S^1$ and mirror symmetry," *Phys. Rev.* **D91** (2015) 105023, arXiv:1408.3371 [hep-th].

Acknowledgments

First of all, I gratefully thank my advisor Yutaka Hosotani for many fruitful discussions on mathematical structures of quantum field theory and for his encouragement throughout 5 years. I think I could not have achieved any meaningful results without his kind advice. I would like to thank Koji Hashimoto for encouraging me and advising me on how to give talks. Thanks to that I have been blessed with many opportunities for discussions.

I would like to thank Satoshi Yamaguchi for talking with me every day. I have learned from him that it is extremely useful to consider problems as simply as possible. I would like to thank Norihiro Iizuka for collaborating with me. His arguments based on physical intuitions were a great shock for me. I learned from him the importance of the physically motivated idea. I would also like to thank Kensuke Kobayashi for reading this thesis carefully from the experimentalist point of view and pointing out some issues to be resolved. As well, I would like to thank Hironori Mori and Takeshi Morita for the past collaborations which became basis of this thesis. Without their great contributions, I could not have accomplished this work and this thesis.

I am also grateful to members of the Korea Institute for Advanced Study Kimyeong Lee, Sung-Soo Kim, Sang-Woo Kim, Futoshi Yagi, Hyun Seok Yang, and Yutaka Yoshida; members of Hanyang University Shigenori Seki, Yunseok Seo, and Sang-Jin Sin; members of Seoul National University Seok Kim, Sangmin Lee, and Daisuke Yokoyama; members of National Taiwan University Heng-Yu Chen, Toshiaki Fujimori, Kazuo Hosomichi, Takeo Inami, and Hiroaki Nakajima; members of the Tokyo Institute of Technology Yosuke Imamura, Katsushi Ito, Tetuuji Kimura, and Hiroki Matsuno; members of Rikkyo University Tohru Eguchi, Chika Hasegawa, and Yasuaki Hikida; members of the Riken String Group Taro Kimura, Shunji Matsuura, Toshifumi Noumi, Noriaki Ogawa, Masato Taki, Shingo Torii, and Yasuhiro Yamaguchi; and Professor of AIMR at Tohoku University Motoko Kotani, for offering me opportunities to give talks on our results; and Dongmin Gang, Yu Nakayama, Sara Pasquetti, and Yuji Sugawara for variable comments.

Finally, I would like to thank Springer's publishing editors, Hisako Niko, Akiyuki Tokuno and other members for many helps on preparation of this publication.

Contents

Chapter 1
Introduction and Summary

Quantum Field Theory (QFT) has been a useful and fundamental tool for studying physics described by large number degrees of freedom, e.g. particle physics, condensed matter physics. In particle physics, theory is typically described by Lagrangian with *Poincaré symmetry* generated by translations and rotations in order to make it compatible with special relativity.[1] One of the generalizations of the Poincaré symmetry, supersymmetry (SUSY), was discovered in 1971 in the context of string theory [2–4]. After that, it was applied to the usual QFT in [5, 6]. The study of SUSY gauge theories has been providing many interesting results including various non-perturbative effects and un-expected relationships between physics and mathematics since 1990's [7–9]. SUSY has generator $\hat{Q}$ with fermionic statistics. One can show that the SUSY algebra is a unique extension of the Poincaré algebra under the existence of a non-trivial S-matrix [10]. Without this condition, there is another extension of the Poincaré symmetry called *Conformal symmetry* generated by translations, rotations, dilatation and conformal boosts. The Conformal symmetry naturally emerges at IR fixed points of *renormalization group flow* [11]. Around each IR fixed point, there is no scale, and this scale invariance enhances to the Conformal symmetry in many cases. See for example [12]. Once we start with supersymmetric UV Lagrangian and flow it to IR regime with preserving supersymmetry, the symmetry of the IR theory is expected to be enhanced to *Superconformal symmetry*. The possible superconformal algebras are classified in [13], and according to it, we can define superconformal theories only within two, three, four, five and six dimensions. Two dimension is in a special case because the algebra becomes infinite dimensional one, so three dimension is the lowest dimension with the *finite dimensional* superconformal algebra, and we focus on the three-dimensional SUSY QFTs from now on.

[1] See [1] for an explanation.

A. Tanaka, *Superconformal Index on* $\mathrm{RP}^2 \times \mathrm{S}^1$ *and 3D Mirror Symmetry*,
Springer Theses, DOI 10.1007/978-981-10-1398-0_1

SUSY QFTs in three dimension have many interesting features. Our main interest is a non-trivial duality among three-dimensional SUSY QFTs, called *three-dimensional mirror symmetry*. It was originally proposed in [14] for $\mathcal{N} = 4$ supersymmetric case, and in [15] for $\mathcal{N} = 2$ supersymmetric case. The simplest case of the duality is an equivalence between the moduli space[2] for *Supersymmetric Quantum ElectroDynamics* (SQED) and the moduli space for a SUSY matter theory called *XYZ-model*. A branch of the moduli space of SQED, so-called Coulomb branch is deformed by the quantum effect [16] but the conjectured dual moduli space branch, called Higgs branch is not because of the non-renormalization theorem [17, 18]. In other words, quantum effect on one side is realized by classical effect on the other side. This proposal is reformulated in the context of the string theory [19, 20], and three-dimensional mirror symmetry was explained as one of the consequences of the $SL(2, Z)$ duality in type IIB superstring theory. In addition to it, this proposal has been checked by utilizing the parity anomaly [15]. It is an analog of 't Hooft anomaly matching condition in four-dimensional duality [7].

There is another way to see an evidence of three-dimensional mirror symmetry. For example, the following equality is expected.

$$Z_{\text{XYZ}} = Z_{\text{SQED}}, \tag{1.1}$$

where Z represents the partition function for each theory. At a first glance, the exact check for (1.1) looks difficult because of the existence of the interaction. Recently, however, so-called *supersymmetric localization techniques* have been developed within 2, 3, 4, 5 dimensional SUSY QFTs.[3] The techniques provide us a way to perform path integral calculations exactly even there are interactions. One of the interesting features for these developments is that the techniques can be applied to the theories on a *curved space*. The curved space, called *manifold* in mathematics, is not arbitrary because we should have a simple structure on the manifold in order to define supersymmetry consistently. In three-dimension, the structure has been identified to so-called *almost integrable contact structure* [40], and the exact calculations were performed on manifolds with such structure, product space $S^2 \times S^1$ [28, 41–43], $D^2 \times S^1$ [30, 33], three sphere S^3 [27, 44–51] and its orbifold S^3/Z_p [32]. In each case, the equality (1.1) has been verified by using mathematical formulas.[4] By the way, the supersymmetric partition functions on $M^2 \times S^1$ where M^2 is a two-dimensional manifold is known to be equivalent to the following object

$$\mathcal{I}^{M^2}_{\text{Theory}}(x, \alpha_a) = \text{Tr}_{\mathcal{H}(M^2)}\Big((-1)^{\hat{F}} x'^{\{Q,Q^\dagger\}} x^{\hat{H}+\hat{j}_3} \prod_a \alpha_a^{\hat{f}_a}\Big), \tag{1.2}$$

[2]It corresponds to the space of possible vacuum expectation values.

[3]One can find results of the localization techniques for two-dimensional QFTs in [21–26], for three-dimensional QFTs in [27–33] for four-dimensional QFTs in [34–36] for five-dimensional QFTs in [37–39].

[4]The check or proof of the equality (1.1) by utilizing supersymmetric partition function on $S^2 \times S^1$, $D^2 \times S^1$, S^3 and S^3/Z_p can be found in [42, 43, 52], [30], [44] and [32] respectively.

called *SuperConformal Index* (SCI). $\hat{j}_3$ and $\hat{f}_a$ are an orbital angular momentum and flavor charges respectively. As reviewed in Chap. 3, this quantity turns to be an analog of usual thermal partition function. Therefore, the following equality which is the counterpart of (1.1) is expected to be satisfied:

$$\mathcal{I}^{M^2}_{\text{XYZ}}(x, \alpha) = \mathcal{I}^{M^2}_{\text{SQED}}\left(x, \alpha^{-1}\right). \tag{1.3}$$

As explained in Chap. 4, thanks to the localization techniques, the structure of exact SCI on $S^2 \times S^1$ for SQED is realized by a summation over the Dirac monopoles labelled by $B \in Z$. As reviewed in Appendix C.1, we have to combine these contributions and utilize fancy mathematical formulas, *Ramanujan's summation formula* and *quantum binomial formula*, in order to deform its infinite summation to the SCI of XYZ-model:

$$\mathcal{I}^{S^2}_{\text{XYZ}}(x, \alpha) \xleftarrow[\text{Ramanujan's summation formula}]{\text{quantum binomial formula +}} \mathcal{I}^{S^2}_{\text{SQED}}\left(x, \alpha^{-1}\right). \tag{1.4}$$

This proof was originally performed in [52], and it provides an explicit evidence for the three-dimensional mirror symmetry.

Our main results We get the following new results.

- We define a new SCI by using $M^2 = RP^2$ in (1.2).
- We derive formulas for the SCI based on localization for $U(1)$ gauge theories.
- We observe the equality (1.3) and prove it in our context.

RP^2 is called *real projective plane*. Topologically, one can construct this curved surface by combining the *Möbius strip* and the hemisphere D^2 along the boundary. RP^2 is not isomorphic to neither S^2 nor D^2. RP^2 is an example for unorientable manifold, and the field theory on it sounds somewhat exotic in usual sense. We define SUSY gauge theories on $RP^2 \times S^1$ by introducing sets of supersymmetric parity condition on $S^2 \times S^1$. The SCI for gauge theory on $RP^2 \times S^1$ consists of a summation over contributions of +holonomy sector and −holonomy sector, and there is no infinitely many terms but just 2 terms, and differ from the SCI on $S^2 \times S^1$. The equality (1.3) is checked numerically in Chap. 6, and we show its exact proof by using *quantum binomial formula* and unnamed formula (7.1) in Appendix C.

$$\mathcal{I}^{RP^2}_{\text{XYZ}}(x, \alpha) \xleftarrow[\text{un-named formula (7.1)}]{\text{quantum binomial formula +}} \mathcal{I}^{RP^2}_{\text{SQED}}\left(x, \alpha^{-1}\right). \tag{1.5}$$

Compared with the proof of (1.4), we can observe that the agreement in (1.5) is guaranteed not by the Ramanujan's formula but another, un-named formula (7.1). We can easily understand its difference because there is no Dirac monopole[5] on RP^2 but $\pm$ holonomies as noted above. The use of the un-named formula (7.1) is an algebraic representation of the $\pm$ holonomies. As one can see, the use of quantum

[5]Precisely speaking, similar object exists even on RP^2 [53]. We will comment on it in the conclusion.

binomial formula is in common. This is also easy to understand because as a common factor, we have Wilson line phase along the thermal S^1. The use of quantum binomial formula is, therefore, an algebraic representation of the Wilson line phase along the thermal S^1.

The organization of this paper is as follows. In Chap. 2, we review some basics of the Quantum Mechanics (QM). This chapter is important because we calculate SCI (1.2) by utilizing the method in Chap. 2. In Chap. 3, we summarize some basic facts on the three-dimensional $\mathcal{N} = 2$ supersymmetry and review the supersymmetric localization techniques. In Chap. 4, we review the exact calculation for the SCI with $M^2 = S^2$ by localization method from the many-body QM point of view. And in Chap. 5, we turn to the calculation with $M^2 = RP^2$ and get new results. Finally, in Chap. 6, we check the simplest three-dimensional mirror symmetry, equivalence between XYZ-model and SQED numerically. If one wants to know how to prove it analytically, see Appendix C. In Chap. 7, we summarize this thesis and comment on new results beyond this thesis.

References

1. S. Weinberg, in *The Quantum Theory of Fields*, vol. 1, The Quantum Theory of Fields 3 Volume Hardback Set. (Cambridge University Press, Cambridge, 1995)
2. P. Ramond, Dual theory for free fermions. Phys. Rev. D **3**, 2415–2418 (1971)
3. A. Neveu, J. Schwarz, Factorizable dual model of pions. Nucl. Phys. B **31**, 86–112 (1971)
4. J.-L. Gervais, B. Sakita, Field theory interpretation of supergauges in dual models. Nucl. Phys. B **34**, 632–639 (1971)
5. J. Wess, B. Zumino, Supergauge transformations in four-dimensions. Nucl. Phys. B **70**, 39–50 (1974)
6. J. Wess, B. Zumino, A lagrangian model invariant under supergauge transformations. Phys. Lett. B **49**, 52 (1974)
7. N. Seiberg, Exact results on the space of vacua of four-dimensional SUSY gauge theories. Phys. Rev. **D49** 6857–6863 (1994). arXiv:hep-th/9402044 [hep-th]
8. N. Seiberg, E. Witten, Electric - magnetic duality, monopole condensation, and confinement in N = 2 supersymmetric Yang-Mills theory. Nucl. Phys. **B426** 19–52 (1994). arXiv:hep-th/9407087 [hep-th]
9. N.A. Nekrasov, Seiberg-Witten prepotential from instanton counting. Adv. Theor. Math. Phys. **7** 831–864 (2004) arXiv:hep-th/0206161 [hep-th]
10. R. Haag, J.T. Lopuszanski, M. Sohnius, All possible generators of supersymmetries of the s matrix. Nucl. Phys. B **88**, 257 (1975)
11. K. Wilson, J.B. Kogut, The renormalization group and the epsilon expansion. Phys. Rept. **12**, 75–200 (1974)
12. Y. Nakayama, A lecture note on scale invariance vs conformal invariance. arXiv:1302.0884 [hep-th]
13. W. Nahm, Supersymmetries and their representations. Nucl. Phys. B **135**, 149 (1978)
14. K.A. Intriligator, N. Seiberg, Mirror symmetry in three-dimensional gauge theories. Phys. Lett. **B387** 513–519 (1996). arXiv:hep-th/9607207 [hep-th]
15. O. Aharony, A. Hanany, K.A. Intriligator, N. Seiberg, M. Strassler, Aspects of N = 2 supersymmetric gauge theories in three-dimensions. Nucl. Phys. **B499** 67–99 (1997). arXiv:hep-th/9703110 [hep-th]

16. I. Affleck, J.A. Harvey, E. Witten, Instantons and (Super)symmetry Breaking in (2+1)-dimensions. Nucl. Phys. B **206**, 413 (1982)
17. N. Seiberg, Naturalness versus supersymmetric nonrenormalization theorems. Phys. Lett. **B318** 469–475 (1993). arXiv:hep-ph/9309335 [hep-ph]
18. P.C. Argyres, M.R. Plesser, N. Seiberg, The Moduli space of vacua of N=2 SUSY QCD and duality in N = 1 SUSY QCD. Nucl. Phys. **B471** 159–194 (1996). arXiv:hep-th/9603042 [hep-th]
19. A. Hanany, E. Witten, Type IIB superstrings, BPS monopoles, and three-dimensional gauge dynamics. Nucl. Phys. **B492** 152–190 (1997). arXiv:hep-th/9611230 [hep-th]
20. J. de Boer, K. Hori, Y. Oz, Z. Yin, Branes and mirror symmetry in N = 2 supersymmetric gauge theories in three-dimensions. Nucl. Phys. **B502** 107–124 (1997). arXiv:hep-th/9702154 [hep-th]
21. N. Doroud, J. Gomis, B. Le Floch, S. Lee, Exact results in D = 2 supersymmetric gauge theories. J. High Energ. Phys. **1305**, 093 (2013). arXiv:1206.2606 [hep-th]
22. F. Benini, S. Cremonesi, Partition functions of N = (2,2) gauge theories on S^2 and vortices. arXiv:1206.2356 [hep-th]
23. H. Kim, S. Lee, P. Yi, Exact partition functions on RP^2 and orientifolds. J. High Energ. Phys. **1402**, 103 (2014). arXiv:1310.4505 [hep-th]
24. F. Benini, R. Eager, K. Hori, Y. Tachikawa, Elliptic genera of two-dimensional N = 2 gauge theories with rank-one gauge groups. Lett. Math. Phys. **104**, 465–493 (2014). arXiv:1305.0533 [hep-th]
25. D. Honda, T. Okuda, Exact results for boundaries and domain walls in 2d supersymmetric theories. arXiv:1308.2217 [hep-th]
26. K. Hori, M. Romo, Exact results in two-dimensional (2,2) Supersymmetric gauge theories with boundary. arXiv:1308.2438 [hep-th]
27. A. Kapustin, B. Willett, I. Yaakov, Exact results for wilson loops in superconformal chern-simons theories with matter. J. High Energ. Phys. **1003**, 089 (2010). arXiv:0909.4559 [hep-th]
28. S. Kim, The Complete superconformal index for N = 6 Chern-Simons theory. Nucl. Phys. B **821**, 241–284 (2009). arXiv:0903.4172 [hep-th]
29. A. Tanaka, H. Mori, T. Morita, Superconformal index on $RP^2 \times S^1$ and mirror symmetry. Phys. Rev. D **91**, 105023 (2015). arXiv:1408.3371 [hep-th]
30. Y. Yoshida, K. Sugiyama, Localization of 3d $\mathcal{N} = \in$ supersymmetric theories on $S^1 \times D^2$. arXiv:1409.6713 [hep-th]
31. J. Källén, Cohomological localization of Chern-Simons theory. J. High Energy Phys. **1108** 008 (2011). arXiv:1104.5353 [hep-th]
32. Y. Imamura, D. Yokoyama, S^3/Z_n partition function and dualities. J. High Energy Phys. **1211**, 122 (2012). arXiv:1208.1404 [hep-th]
33. S. Sugishita, S. Terashima, Exact results in supersymmetric field theories on manifolds with boundaries. J. High Energy Phys. **1311**, 021 (2013). arXiv:1308.1973 [hep-th]
34. V. Pestun, Localization of gauge theory on a four-sphere and supersymmetric Wilson loops. Commun. Math. Phys. **313**, 71–129 (2012). arXiv:0712.2824 [hep-th]
35. S. Terashima, A localization computation in confining phase. arXiv:1410.3630 [hep-th]
36. T. Nishioka, I. Yaakov, Generalized indices for $\mathcal{N} = 1$ theories in four-dimensions. J. High Energy Phys. **1412**, 150 (2014). arXiv:1407.8520 [hep-th]
37. J. Kallen, J. Qiu, M. Zabzine, The perturbative partition function of supersymmetric 5D Yang-Mills theory with matter on the five-sphere. J. High Energy Phys. **1208**, 157 (2012). arXiv:1206.6008 [hep-th]
38. Y. Imamura, Perturbative partition function for squashed S^5. arXiv:1210.6308 [hep-th]
39. H.-C. Kim, S. Kim, S.-S. Kim, K. Lee, The general M5-brane superconformal index. arXiv:1307.7660
40. C. Closset, T.T. Dumitrescu, G. Festuccia, Z. Komargodski, Supersymmetric field theories on three-manifolds. J. High Energy Phys. **1305**, 017 (2013). arXiv:1212.3388 [hep-th]
41. D. Gang, Chern-Simons theory on L(p,q) lens spaces and localization. arXiv:0912.4664 [hep-th]

42. Y. Imamura, S. Yokoyama, Index for three dimensional superconformal field theories with general R-charge assignments. J. High Energy Phys. **1104**, 007 (2011). arXiv:1101.0557 [hep-th]
43. A. Kapustin, B. Willett, Generalized superconformal index for three dimensional field theories. arXiv:1106.2484 [hep-th]
44. N. Hama, K. Hosomichi, S. Lee, Notes on susy gauge theories on three-sphere. J. High Energy Phys. **1103**, 127 (2011). arXiv:1012.3512 [hep-th]
45. D.L. Jafferis, The exact superconformal R-Symmetry extremizes Z. J. High Energy Phys. **1205**, 159 (2012). arXiv:1012.3210 [hep-th]
46. N. Hama, K. Hosomichi, S. Lee, SUSY Gauge theories on squashed three-spheres. J. High Energy Phys. **1105**, 014 (2011). arXiv:1102.4716 [hep-th]
47. Y. Imamura, D. Yokoyama, N = 2 supersymmetric theories on squashed three-sphere. Phys. Rev. D **85**, 025015 (2012). arXiv:1109.4734 [hep-th]
48. L.F. Alday, D. Martelli, P. Richmond, J. Sparks, Localization on three-manifolds. arXiv:1307.6848 [hep-th]
49. J. Nian, Localization of supersymmetric chern-simons-matter theory on a squashed S^3 with $SU(2) \times U(1)$ Isometry. arXiv:1309.3266 [hep-th]
50. A. Tanaka, Localization on round sphere revisited. J. High Energy Phys. **1311**, 103 (2013). arXiv:1309.4992 [hep-th]
51. M. Fujitsuka, M. Honda, Y. Yoshida, Higgs branch localization of 3d $\mathcal{N} = 2$ theories. PTEP **12** 123B02 (2014). arXiv:1312.3627 [hep-th]
52. C. Krattenthaler, V. Spiridonov, G. Vartanov, Superconformal indices of three-dimensional theories related by mirror symmetry. J. High Energy Phys. **1106**, 008 (2011). arXiv:1103.4075 [hep-th]
53. A. Tanaka, H. Mori, T. Morita, Abelian 3d mirror symmetry on $\mathrm{RP}^2 \times \mathrm{S}^1$ with $N_f = 1$. J. High Energy Phys. **09**, 154 (2015). arXiv:1505.07539 [hep-th]

Chapter 2
Preliminary—Quantum Mechanics

This chapter is a preliminary chapter for forthcoming discussions. First, we briefly review representation theories for boson and fermion. Second, we turn to consider partition function

$$Z = \mathrm{Tr}(e^{-\beta\hat{H}}). \tag{2.1}$$

Third, we generalize it by turning on an insertion of $(-1)^{\hat{F}}$ into the trace:

$$I = \mathrm{Tr}\Big((-1)^{\hat{F}} e^{-\beta\hat{H}}\Big). \tag{2.2}$$

This quantity is called *Witten index*, a prototype of the superconformal index in Chaps. 3–5. $\hat{F}$ is fermion number operator which counts the number of fermionic excitations. In the last section, we generalize it and the generalized index gives the basis for Chap. 3.

2.1 Representation Theory

We briefly review the basics of boson and fermion in QM. We emphasis the relationship between operator formalism and path integral formalism for later use.

2.1.1 Boson

Classical prescription Bosonic Lagrangian typically takes the following form:

A. Tanaka, *Superconformal Index on* $\mathrm{RP}^2 \times \mathrm{S}^1$ *and 3D Mirror Symmetry*,
Springer Theses, DOI 10.1007/978-981-10-1398-0_2

$$L_b = \frac{1}{2}\dot{x}^2 - V(x). \tag{2.3}$$

As a next step, we define the conjugate momentum of x by

$$p = \frac{\partial L_b}{\partial \dot{x}}. \tag{2.4}$$

Then, the Hamiltonian is defined by the Legendre transformation of L_b:

$$\begin{aligned} H_b &= p\dot{x} - L_b \\ &= \frac{1}{2}p^2 + V(x). \end{aligned} \tag{2.5}$$

Canonical quantization We start with the representation of the bosonic algebra, i.e. Heisenberg algebra:

$$[\hat{p}, \hat{x}] = -i, \tag{2.6}$$

where $\hat{p}$ and $\hat{x}$ are momentum and position operators respectively. In principle, we do not need to stick on the definition of $\pm$ sign in (2.6) if we treat it in self consistent way [1]. As a basis of the Hilbert space, we can take

$$|x\rangle \quad \text{or} \quad |p\rangle. \tag{2.7}$$

These states are defined by

$$\hat{x}|x\rangle = x|x\rangle, \quad \int_{-\infty}^{+\infty} dx\ |x\rangle\langle x| = 1, \tag{2.8}$$

$$\hat{p}|p\rangle = p|p\rangle, \quad \int_{-\infty}^{+\infty} dp\ |p\rangle\langle p| = 1. \tag{2.9}$$

There are two important facts. First fact is that $e^{-i\hat{p}a}$ generates translation of $|x\rangle$:

$$e^{-i\hat{p}a}|x\rangle = |x + a\rangle. \tag{2.10}$$

Second fact is that the explicit form of the inner product becomes as follows.[1]

$$\langle p|x\rangle = \frac{1}{\sqrt{2\pi}}e^{-ipx}. \tag{2.11}$$

[1]The simplest way to derive this relation is to use the differential equation. For example,

The constant of integration, $\frac{1}{\sqrt{2\pi}}$, is determined by requiring the orthonormality condition $\langle x'|x\rangle = \delta(x - x')$.

2.1.2 Fermion

Classical prescription Fermionic Lagrangian typically takes the following form:

$$L_f = i\psi_+\dot{\psi}_- - V(\psi_\pm). \tag{2.12}$$

Here we treat ψ_+, ψ_- as independent Grassmann numbers:

$$\psi_+^2 = 0, \quad \psi_-^2 = 0, \quad \psi_+\psi_- = -\psi_-\psi_+. \tag{2.13}$$

The left[2] conjugate momentum of ψ_- is defined by

$$\Pi_- = \frac{\partial}{\partial\dot{\psi}_-}L_f. \tag{2.14}$$

The Hamiltonian is defined by the Legendre transformation of L_f:

$$\begin{aligned} H_f &= \Pi_-\dot{\psi}_- - L_f \\ &= V(\psi_\pm). \end{aligned} \tag{2.15}$$

Canonical quantization We start with the representation of the fermionic algebra, i.e. Clifford algebra[3]:

$$\{\hat{\psi}_+, \hat{\psi}_-\} = +1. \tag{2.16}$$

(Footnote 1 continued)

$$\begin{aligned} \frac{\partial}{\partial x}\langle p|x\rangle &= \lim_{a\to 0}\frac{\langle p|x+a\rangle - \langle p|x\rangle}{a} \\ &= \lim_{a\to 0}\frac{\langle p|e^{-i\hat{p}a}|x\rangle - \langle p|x\rangle}{a} \\ &= \lim_{a\to 0}\frac{e^{-ipa}\langle p|x\rangle - \langle p|x\rangle}{a} \\ &= -ip\langle p|x\rangle. \end{aligned}$$

[2]Because of the fermionic natures in (2.13), we have to be careful with the order of ψ_+ and ψ_-.

[3]In order to derive this relation from the usual canonical quantization method, considering Poisson bracket is not enough. Instead of it, Dirac bracket is necessary.

In contrast to the bosonic case, the sign in (2.16) is important to get the unitary representation [1]. As an orthonormal basis of the Hilbert space, we can take

$$\left\{ |0\rangle, |1\rangle \right\}. \tag{2.17}$$

The states $|0\rangle, |1\rangle$ are defined by

$$\begin{aligned} &\hat{\psi}_-|0\rangle = 0, \quad \hat{\psi}_+|0\rangle = |1\rangle, \\ &\hat{\psi}_-|1\rangle = |0\rangle, \quad \hat{\psi}_+|1\rangle = 0, \\ &|0\rangle\langle 0| + |1\rangle\langle 1| = 1. \end{aligned} \tag{2.18}$$

One can regard $|0\rangle$ as a hole-state, and $|1\rangle$ as an occupied state. We cannot make $|2\rangle := \psi_+|1\rangle$ because it is automatically zero. This is an algebraic representation of the famous Pauli exclusion principle.

Coherent state basis In later discussions, we consider the path integral formalism. In order to derive it, there is a more useful basis than the basis in (2.17), the coherent state basis [2]:

$$|\Psi\rangle = e^{-\Psi\hat{\psi}_+}|0\rangle, \quad \langle\Psi| = \langle 0|e^{\Psi\hat{\psi}_-}. \tag{2.19}$$

We should take Ψ as a Grassmann valuable, therefore $\Psi^2 = 0$ and we get

$$|\Psi\rangle = (1 - \Psi\hat{\psi}_+)|0\rangle. \tag{2.20}$$

These states satisfy the following relations.

$$\hat{\psi}_-|\Psi\rangle = \Psi|\Psi\rangle, \quad \langle\Psi|\hat{\psi}_+ = \langle\Psi|\Psi. \tag{2.21}$$

After a direct calculation, one can get the inner product formula

$$\langle\Psi_+|\Psi_-\rangle = e^{\Psi_+\Psi_-} \tag{2.22}$$

and the complete relation

$$\int d\Psi_+ d\Psi_- |\Psi_-\rangle e^{-\Psi_+\Psi_-} \langle\Psi_+| = 1. \tag{2.23}$$

2.2 Partition Function

One of the most important objects in QM is the partition function:

$$Z = \mathrm{Tr}(e^{-\beta\hat{H}}). \tag{2.24}$$

It contains all informations of the energy spectra because we can extract each energy eigenvalue by taking following procedure [3, 4][4]:

1. Taking $\beta \to \infty$ of Z, then Z behaves $e^{-\beta E_0}$ where E_0 is the ground state energy.
2. Subtracting $e^{-\beta E_0}$ from Z, and rename it Z_1, and

   taking $\beta \to \infty$ of Z_1, then Z_1, behaves $e^{-\beta E_1}$ where E_1 is the 1st exited state energy.
3. Repeating this procedure.

2.2.1 *Boson Sector*

Partition function of the bosonic degrees of freedom is described by the Hamiltonian operator defined from (2.5):

$$\hat{H} = \hat{H}_b, \quad \hat{H}_b = \frac{1}{2}\hat{p}^2 + V(\hat{x}). \tag{2.25}$$

Harmonic oscillator The simplest example is

$$V(\hat{x}) = \frac{1}{2}\omega^2\hat{x}^2. \tag{2.26}$$

In this case, as well known, once we define $\hat{a}$ and $\hat{a}^\dagger$ [5] so that

$$\hat{H}_b = \omega\left(\hat{a}^\dagger\hat{a} + \frac{1}{2}\right), \tag{2.27}$$

and by constructing a basis

$$\left\{|0\rangle, |1\rangle, |2\rangle, \dots\right\}, \quad \hat{a}|n\rangle = \sqrt{n}|n-1\rangle, \quad \hat{a}^\dagger|n\rangle = \sqrt{n+1}|n+1\rangle, \tag{2.28}$$

[4]This is valid if there is no degeneracy.

then, we can diagonalize the Hamiltonian: $\hat{H}_b|n\rangle = \omega(n+\frac{1}{2})|n\rangle$. By using this basis, the partition function can be computed by utilizing the formula of power series

$$\begin{aligned}\mathrm{Tr}(e^{-\beta\hat{H}_b}) &= \sum_{n=0}^{\infty} e^{-\beta\omega(n+\frac{1}{2})} \\ &= \frac{e^{-\frac{\beta\omega}{2}}}{1-e^{-\beta\omega}} \\ &= \frac{1}{2\sinh\frac{\beta\omega}{2}}.\end{aligned} \tag{2.29}$$

The zero energy which corresponds to $n = 0$ is often called Casimir energy.

Path integral formalism By Inserting the complete set (2.8) and (2.9) into the trace in (2.24), we can re-express it as

$$\begin{aligned}Z_b &= \int_{x(0)=x(\beta)} \left(\prod_{t\in[0,\beta]} dx(t)\frac{dp(t)}{2\pi}\right) e^{-\int_0^\beta dt\left(ip\dot{x}+\frac{1}{2}p^2+V(x)\right)} \\ &= \int_{x(0)=x(\beta)} \left(\prod_{t\in[0,\beta]} \frac{dx(t)}{\sqrt{2\pi}}\right) e^{-\int_0^\beta dt\left(\frac{-1}{2}x\partial_t^2 x+V(x)\right)}.\end{aligned} \tag{2.30}$$

Path integral description of harmonic oscillator We have the following action

$$-\int_0^\beta dt\Big(\frac{-1}{2}x\partial_t^2 x + V(x)\Big) = -\frac{1}{2}\int_0^\beta dt\ x(-\partial_t^2+\omega^2)x. \tag{2.31}$$

Thanks to the Gaussian integral formula in (A.13), we get

$$Z_b = \frac{1}{\sqrt{\det_{x(0)=x(\beta)}(-\partial_t^2+\omega^2)}}. \tag{2.32}$$

The "matrix" ∂_t's eigenvectors are $x_n(t) = e^{\frac{2\pi i}{\beta}nt},\ n \in Z$ because

$$\partial_t x_n(t) = \frac{2\pi i}{\beta} n x_n(t). \tag{2.33}$$

Therefore, we get the following representation of the determinant.

$$\det_{x(0)=x(\beta)}(-\partial_t^2+\omega^2) = \prod_{n=-\infty}^{\infty}\left(\frac{(2\pi)^2}{\beta^2}n^2+\omega^2\right)$$

$$= \omega^2 \Big[\prod_{n=1}^{\infty} \Big(\frac{(2\pi)^2}{\beta^2} n^2 + \omega^2 \Big) \Big]^2$$

$$= \Big[\prod_{n=1}^{\infty} \frac{2\pi}{\beta} n \Big]^4 \times \omega^2 \prod_{n=1}^{\infty} \Big(1 + \frac{(\beta\omega)^2}{(2\pi n)^2} \Big)^2 . \tag{2.34}$$

Obviously, the first factor diverges. We regularize it by using zeta-function regularization. (See Appendix A for $\zeta(0)$, $\zeta'(0)$ values' derivation.):

$$\Big[\prod_{n=1}^{\infty} \frac{2\pi}{\beta} n \Big]^4 = \exp\Big(4 \sum_{n=1}^{\infty} \log \frac{2\pi}{\beta} n \Big) \to \exp\Big(4 \Big[-\zeta'(0) - \zeta(0) \log \frac{\beta}{2\pi} \Big] \Big)$$
$$= \exp\Big(4 \Big[-(-\frac{1}{2} \log 2\pi) - (-\frac{1}{2}) \log \frac{\beta}{2\pi} \Big] \Big)$$
$$= \beta^2 . \tag{2.35}$$

Then, by using the infinite product formula (A.1), we get

$$(2.34) = \Big[(\beta\omega) \prod_{n=1}^{\infty} \Big(1 + \frac{(\beta\omega)^2}{(2\pi n)^2} \Big) \Big]^2 = \Big[2 \sinh \frac{\beta\omega}{2} \Big]^2 . \tag{2.36}$$

It reproduces the result (2.29):

$$Z_b = \frac{1}{\sqrt{\det_{x(0)=x(\beta)} \left(-\partial_t^2 + \omega^2 \right)}} = \frac{1}{2 \sinh \frac{\beta\omega}{2}} . \tag{2.37}$$

2.2.2 *Fermion Sector*

Partition function of the fermionic degrees of freedom is described by the Hamiltonian operator defined from (2.15):

$$\hat{H} = \hat{H}_f , \quad \hat{H}_f = V(\hat{\psi}_\pm). \tag{2.38}$$

Harmonic oscillator The simplest example is

$$V(\hat{\psi}_\pm) = \omega \left(\hat{\psi}_+ \hat{\psi}_- - \frac{1}{2} \right) . \tag{2.39}$$

Then, the basis (2.17) diagonalizes this Hamiltonian:

$$\hat{H}_f|n\rangle = \omega\left(n - \frac{1}{2}\right)|n\rangle, \quad n = 0, 1. \tag{2.40}$$

The partition function is, therefore,

$$\begin{aligned}\mathrm{Tr}(e^{-\beta\hat{H}_f}) &= \sum_{n=0}^{1} e^{-\beta\omega(n-\frac{1}{2})} \\ &= e^{\frac{\beta\omega}{2}} + e^{-\frac{\beta\omega}{2}} \\ &= 2\cosh\frac{\beta\omega}{2}.\end{aligned} \tag{2.41}$$

There are two important differences compared with the bosonic harmonic oscillator.

- The absolute value of Casimir energy is same but the sign is different.
- cosh function appears unlike the sinh in bosonic case.

As we will see later, if we insert $(-1)^{\hat{\psi}_+\hat{\psi}_-}$ into the trace, we get sinh not cosh.

Path integral formalism When we derive fermion's path integral representation of the partition function, we have to be careful with the periodicity as described below. First, we re-express Tr in the partition function with coherent basis in (2.19):

$$\begin{aligned}Z_f &= \mathrm{Tr}(e^{-\beta\hat{H}_f}) \\ &= \int d\Psi_+ d\Psi_-\ e^{\Psi_+\Psi_-}\langle\Psi_+|e^{-\beta V(\hat{\psi}_+,\hat{\psi}_-)}|\Psi_-\rangle.\end{aligned} \tag{2.42}$$

Second, we divide β into N pieces: $\epsilon = \frac{\beta}{\mathrm{N}}$, say N=2,

$$\begin{aligned}(2.42) &= \int d\Psi_+ d\Psi_- \int d\Lambda_+ d\Lambda_-\ e^{\Psi_+\Psi_-}\langle\Psi_+|e^{-\epsilon V(\hat{\psi}_+,\hat{\psi}_-)}|\Lambda_-\rangle e^{-\Lambda_+\Lambda_-}\langle\Lambda_+|e^{-\epsilon V(\hat{\psi}_+,\hat{\psi}_-)}|\Psi_-\rangle \\ &= \int d\Psi_+ d\Psi_- \int d\Lambda_+ d\Lambda_-\ e^{\Psi_+\Psi_-} e^{-\epsilon V(\Psi_+,\Lambda_-)}\langle\Psi_+|\Lambda_-\rangle e^{-\Lambda_+\Lambda_-}\langle\Lambda_+|\Psi_-\rangle e^{-\epsilon V(\Lambda_+,\Psi_-)} \\ &= \int d\Psi_+ d\Psi_- \int d\Lambda_+ d\Lambda_-\ e^{\Psi_+\Psi_- + \Psi_+\Lambda_- - \Lambda_+\Lambda_- + \Lambda_+\Psi_-} e^{-\epsilon V(\Psi_+,\Lambda_-) - \epsilon V(\Lambda_+,\Psi_-)}.\end{aligned} \tag{2.43}$$

We rename fermionic valuables:

$$\Psi_+ = \Psi_+^2, \quad \Lambda_- = \Psi_-^2, \quad \Lambda_+ = \Psi_+^1, \quad \Psi_- = \Psi_-^1, \tag{2.44}$$

then we get

$$(2.43) = \int d\Psi_+^2 d\Psi_-^2 d\Psi_+^1 d\Psi_-^1\ e^{\Psi_+^2\Psi_-^1 + \Psi_+^2\Psi_-^2 - \Psi_+^1\Psi_-^2 + \Psi_+^1\Psi_-^1 - \epsilon V(\Psi_+^2,\Psi_-^2) - \epsilon V(\Psi_+^1,\Psi_-^1)}. \tag{2.45}$$

Now, we regard each $\Psi_\pm^n$ as $\Psi_\pm(t_n) = \Psi_\pm^n$, where $t_n = \epsilon n$. In this N=2 case,

$$
\begin{aligned}
&\Psi_+^2\Psi_-^1 + \Psi_+^2\Psi_-^2 - \Psi_+^1\Psi_-^2 + \Psi_+^1\Psi_-^1 \\
&= \Psi_+(t_2)\Big(\underbrace{\Psi_-(t_1)}_{\Psi_-(0)+\epsilon\dot{\Psi}_-(0)} + \Psi_-(t_2)\Big) - \Psi_+(t_1)\Big(\underbrace{\Psi_-(t_2)}_{\Psi_-(t_1)+\epsilon\dot{\Psi}_-(t_1)} - \Psi_-(t_1)\Big) \\
&= \Psi_+(t_2)\Big(\epsilon\dot{\Psi}_-(0) + \underbrace{[\Psi_-(0) + \Psi_-(t_2)]}_{\text{we have to make it zero.}}\Big) - \Psi_+(t_1)\Big(\epsilon\dot{\Psi}_-(t_1) + \underbrace{[\Psi_-(t_1) - \Psi_-(t_1)]}_{0}\Big)
\end{aligned}
$$

As we can see above, in order to drop the $\mathcal{O}(\epsilon^0)$ term, we have to take

$$\Psi_-(t_2) = \Psi_-(\beta) = -\Psi_-(0). \tag{2.46}$$

Therefore, corresponding fermionic fields $\Psi_\pm(t)$ are anti-periodic[5] under the translation $t \to t + \beta$. Then, by using

$$\dot{\Psi}(0) = \frac{d}{dt}\Big|_{t=0}\Psi(t) = -\frac{d}{dt}\Big|_{t=0}\Psi(t+\beta) = -\dot{\Psi}(t_2), \tag{2.47}$$

and taking N→ ∞ limit, we arrive at

$$(2.45) = \int_{\Psi_\pm(0)=-\Psi_\pm(\beta)}\left(\prod_{t\in[0,\beta]} d\Psi_+(t)d\Psi_-(t)\right) e^{-\int_0^\beta dt\left(\Psi_+\dot{\Psi}_- + V(\Psi_+,\Psi_-)\right)}. \tag{2.48}$$

Path integral description of harmonic oscillator For harmonic oscillator (2.39),

$$
\begin{aligned}
\mathrm{Tr}(e^{-\beta\hat{H}_f}) &= \int_{\Psi_\pm(0)=-\Psi_\pm(\beta)}\left(\prod_{t\in[0,\beta]} d\Psi_+(t)d\Psi_-(t)\right) e^{-\int_0^\beta dt\ \Psi_+(\partial_t+\omega)\Psi_-} \\
&= \det_{\Psi_\pm(0)=-\Psi_\pm(\beta)}(\partial_t + \omega).
\end{aligned}
\tag{2.49}
$$

We used the Gaussian integral formula for fermionic variables (A.16). In this anti-periodic sector, the eigenvectors of ∂_t are $\psi_n(t) = e^{\frac{2\pi i}{\beta}(n-\frac{1}{2})t}$ with $n \in \mathbb{Z}$. Therefore,

$$
\begin{aligned}
\det_{\Psi_\pm(0)=-\Psi_\pm(\beta)}(\partial_t + \omega) &= \prod_{n=-\infty}^{\infty}\left(\frac{2\pi i}{\beta}(n-\frac{1}{2}) + \omega\right) \\
&= \prod_{n=1}^{\infty}\left(\frac{(2\pi)^2}{\beta^2}(n-\frac{1}{2})^2 + \omega^2\right)
\end{aligned}
$$

[5] We have checked it only with Ψ_-, but we can understand the case for Ψ_+ in similar way.

$$= \left[\prod_{n=1}^{\infty} \frac{2\pi}{\beta}(n - \frac{1}{2}) \right]^2 \times \prod_{n=1}^{\infty} \left(1 + \frac{(\beta\omega)^2}{(2\pi[n - \frac{1}{2}])^2} \right). \tag{2.50}$$

The first factor diverges, so we have to regularize it somehow. One might think that the zeta-function regularization works, however in this case, we should be more careful:

$$\begin{aligned} \left[\prod_{n=1}^{\infty} \frac{2\pi}{\beta}(n - \frac{1}{2}) \right]^2 &= \left[\prod_{n=1}^{\infty} \frac{2\pi}{\beta} n \right]^2 \times \left[\prod_{n=1}^{\infty} \frac{\frac{2\pi}{\beta}(n - \frac{1}{2})}{\frac{2\pi}{\beta} n} \right]^2 \\ &= \left[\prod_{n=1}^{\infty} \frac{2\pi}{\beta} n \right]^2 \times \left[\prod_{n=1}^{\infty} \frac{\frac{\pi}{\beta}(2n - 1)}{\frac{\pi}{\beta}(2n)} \right]^2 \\ &= \underbrace{\left[\prod_{n=1}^{\infty} \frac{2\pi}{\beta} n \right]^2}_{\to \beta} \times \frac{\pi}{\beta} \times \underbrace{\left[\prod_{n=1}^{\infty} \frac{\frac{\pi}{\beta}(2n - 1) \times \frac{\pi}{\beta}(2n + 1)}{\frac{\pi^2}{\beta^2}(2n)^2} \right]}_{\frac{2}{\pi}} \\ &\to 2, \end{aligned} \tag{2.51}$$

where we used Wallis' formula. Another part of (2.50) can be calculated by using infinite product formula for cosh (A.2):

$$\prod_{n=1}^{\infty} \left(1 + \frac{(\beta\omega)^2}{(2\pi[n - \frac{1}{2}])^2} \right) = \cosh \frac{\beta\omega}{2}. \tag{2.52}$$

Gathering all, we recover the result (2.41)

$$\mathrm{Tr}(e^{-\beta \hat{H}_f}) = 2 \cosh \frac{\beta\omega}{2}. \tag{2.53}$$

2.3 Witten Index

As we have reviewed briefly, the partition function of the harmonic oscillator can be calculated easily. However, once we turn on the cubic or more higher interaction, it is difficult to perform the calculation explicitly. In addition to it, the naive zeta-function regularization did not work in the fermionic sector as we have observed in previous page. However, we can overcome such a situation by considering

$$I = \mathrm{Tr}\Big((-1)^{\hat{F}} e^{-\beta \hat{H}} \Big), \quad \text{where } \hat{F} \text{ is a fermion number operator,} \tag{2.54}$$

instead of Z.

Fermion number operator $\hat{F}$ is an operator which counts the number of fermion excitation, 0 or 1. Explicitly, we can write it in our previous notation as

$$\hat{F} = \hat{\psi}_+ \hat{\psi}_- . \tag{2.55}$$

As one can check,

$$(-1)^{\hat{F}} = \begin{cases} +1 \text{ bosonic state} \\ -1 \text{ fermionic state} \end{cases} . \tag{2.56}$$

Therefore, within only bosonic sector, I and Z are identical:

$$I_b = \mathrm{Tr}_b\Big((-1)^{\hat{F}} e^{-\beta \hat{H}_b}\Big) = \mathrm{Tr}_b(e^{-\beta \hat{H}_b}) = Z_b, \tag{2.57}$$

and nothing different happens compared with the partition function. However, the fermion sector's behavior changes drastically.

2.3.1 Fermion Sector

Let us see what happens in the operator formalism first by using the harmonic oscillator example.

Operator formalism We can get I for fermion sector just by inserting $(-1)^n$ into the previous summation in (2.41) as

$$\begin{aligned} \mathrm{Tr}\Big((-1)^{\hat{F}} e^{-\beta \hat{H}_f}\Big) &= \sum_{n=0}^{1} (-1)^n e^{-\beta\omega(n-\frac{1}{2})} \\ &= e^{\frac{\beta\omega}{2}} - e^{-\frac{\beta\omega}{2}} \\ &= 2 \sinh \frac{\beta\omega}{2} . \end{aligned} \tag{2.58}$$

Path integral formalism After a simple calculation, one can verify that

$$\begin{aligned} I &= \mathrm{Tr}\Big((-1)^{\hat{F}} e^{-\beta \hat{H}_f}\Big) \\ &= \int d\Psi_+ d\Psi_- \, e^{-\Psi_+\Psi_-} \langle \Psi_+ | e^{-\beta \hat{H}_f} \Psi_- \rangle . \end{aligned} \tag{2.59}$$

Compering with the partition function (2.42), one can see that the sign of the exponential factor is different. This minus sign makes the fermionic fields $\Psi_\pm(t)$ in the path integral periodic under $t \to t + \beta$. In summary,

$$I_f = \int_{\Psi_\pm(0)=\Psi_\pm(\beta)} \Big(d\Psi_+(t)d\Psi_-(t)\Big) e^{-\int_0^\beta \Big(\Psi_+\partial_t\Psi_- + V(\Psi_+,\Psi_-)\Big)}. \tag{2.60}$$

In this case, we can recover the result (2.58) as follows.

Harminic oscillator

$$\begin{aligned} I_f &= \int_{\Psi_\pm(0)=\Psi_\pm(\beta)} \Big(d\Psi_+(t)d\Psi_-(t)\Big) e^{-\int_0^\beta \Psi_+(\partial_t+\omega)\Psi_-} \\ &= \det_{\Psi_\pm(0)=\Psi_\pm(\beta)} (\partial_t + \omega) \\ &= \prod_{n=-\infty}^{\infty} \left(\frac{2\pi i}{\beta} n + \omega\right) \\ &= \omega \prod_{n=1}^{\infty} \left(\frac{(2\pi n)^2}{\beta^2} + \omega^2\right). \end{aligned} \tag{2.61}$$

The same infinite product in the bosonic partition function (2.34) emerges. Therefore, by repeating zeta-function regularization procedure, we arrive at

$$I_f = 2\sinh\frac{\beta\omega}{2}. \tag{2.62}$$

2.3.2 *Supersymmetric Quantum Mechanics*

What happens when we consider

$$I = \mathrm{Tr}\Big((-1)^{\hat{F}} e^{-\beta\hat{H}}\Big), \tag{2.63}$$

with harmonic oscillator Hamiltonian $\hat{H} = \hat{H}_b + \hat{H}_f$? The answer is extremely simple;

$$\begin{aligned} I &= I_b \times I_f \\ &= Z_b \times I_f \\ &= \frac{1}{2\sinh\frac{\beta\omega}{2}} \times 2\sinh\frac{\beta\omega}{2} \\ &= 1. \end{aligned} \tag{2.64}$$

Note that if we turn on different frequencies ω_b, ω_f for boson and fermion respectively, we get

$$I = \frac{\sinh \frac{\beta\omega_f}{2}}{\sinh \frac{\beta\omega_b}{2}}, \tag{2.65}$$

and it does depend on β. Therefore, the β independence is equivalent to the condition $\omega_b = \omega_f$. It is strongly related to the concept of *supersymmetry*. In other words, the Hamiltonian

$$\hat{H} = \omega\left(\hat{a}^\dagger\hat{a} + \frac{1}{2}\right) + \omega\left(\hat{\psi}_+\hat{\psi}_- - \frac{1}{2}\right) = \omega(\hat{a}^\dagger\hat{a} + \hat{\psi}_+\hat{\psi}_-) \tag{2.66}$$

defines supersymmetric quantum mechanics. The physical meaning is also extremely simple: the state $|0\rangle$ only contributes. This quantity is called Witten index [6]. We can learn other facts of supersymmetry from this extremely simple example by defining

$$\hat{Q} := \sqrt{\omega}\,\hat{a}^\dagger\hat{\psi}_-, \quad \hat{Q}^\dagger := \sqrt{\omega}\,\hat{a}\hat{\psi}_+. \tag{2.67}$$

These operators are called *supercharges* which satisfy the following equation.

$$\hat{H} = \{\hat{Q}, \hat{Q}^\dagger\}. \tag{2.68}$$

By using this expression, the reason for β independence of Witten index becomes clear because the differential of the index with respect to β becomes zero:

$$\begin{aligned} \frac{d}{d\beta}\mathrm{Tr}(-1)^{\hat{F}}e^{-\beta\hat{H}} &= \frac{d}{d\beta}\mathrm{Tr}(-1)^{\hat{F}}e^{-\beta\{\hat{Q},\hat{Q}^\dagger\}} \\ &= -\mathrm{Tr}(-1)^{\hat{F}}(\hat{Q}\hat{Q}^\dagger + \hat{Q}^\dagger\hat{Q})e^{-\beta\{\hat{Q},\hat{Q}^\dagger\}} \\ &= -\mathrm{Tr}(-1)^{\hat{F}}(\hat{Q}\hat{Q}^\dagger - \hat{Q}\hat{Q}^\dagger)e^{-\beta\{\hat{Q},\hat{Q}^\dagger\}} = 0. \end{aligned} \tag{2.69}$$

We can construct a somewhat more non-trivial Hamiltonian (e.g. [6–8]) which contains interaction terms. In such case, Witten index counts the number of degeneracy of ground states, or more technically speaking, it counts the number of *BPS states*.

2.3.3 Generalized Index

In (2.69), we use the following facts:

$$[\hat{H}, \hat{Q}] = [\hat{H}, \hat{Q}^\dagger] = 0. \tag{2.70}$$

It means $\hat{Q}$ and $\hat{Q}^\dagger$ generate symmetry of the system. Suppose there is another generator $\hat{J}$ which commutes with the supercharges:

$$[\hat{Q}, \hat{J}] = 0, \quad [\hat{Q}^\dagger, \hat{J}] = 0, \tag{2.71}$$

then following trace

$$\mathrm{Tr}\Big((-1)^{\hat{F}} e^{-\beta\{\hat{Q},\hat{Q}^\dagger\}} e^{-i\mu\hat{J}}\Big) \tag{2.72}$$

also does not depend on β. In Chap. 3, we introduce the concept of *SuperConformal Index* (SCI). SCI can be regarded such a generalized index. $e^{-i\mu\hat{J}}$ insertion makes $x(t)$ and $\Psi_\pm(t)$ not periodic but "twisted"

$$x(t+\beta) = e^{i\mu J_x} x(t), \quad \Psi_\pm(t+\beta) = e^{i\mu J_\psi} \Psi_\pm(t), \tag{2.73}$$

where J_x, J_ψ are eigenvalues of $\hat{J}$ operator. The reason is as follows. For bosonic degrees of freedom, (2.72) can be expressed

$$\begin{aligned}
\mathrm{Tr}\Big((-1)^{\hat{F}} e^{-\beta\{\hat{Q},\hat{Q}^\dagger\}} e^{-i\mu\hat{J}}\Big) &= \int dx \langle x|(-1)^{\hat{F}} e^{-\beta\{\hat{Q},\hat{Q}^\dagger\}} e^{-i\mu\hat{J}}|x\rangle \\
&= \int dx \langle x|e^{-\beta\hat{H}}|e^{-i\mu J_x} x\rangle \\
&= \int dx dp dx_1 \langle x|e^{-(\beta-\epsilon)\hat{H}}|x_1\rangle \underbrace{\langle x_1|e^{-\epsilon\hat{H}}|p\rangle}_{e^{-\epsilon H(x_1,p)+ipx_1}} \underbrace{\langle p|e^{-i\mu J_x} x\rangle}_{e^{-ipe^{-i\mu J_x}x}},
\end{aligned} \tag{2.74}$$

and at the edge, we have

$$e^{-\epsilon H(x_1,p)+ipx_1-ipe^{-i\mu J_x}x}. \tag{2.75}$$

In order to get rid of $\mathcal{O}(\epsilon^0)$ term,

$$\begin{aligned}
+ipx_1 - ipe^{-i\mu J_x}x &= ip(x_1 - e^{-i\mu J_x}x) \\
&= ip\Big(x(t=\epsilon) - e^{-i\mu J_x}x(t=\beta)\Big) \\
&= ip\Big(\epsilon\dot{x}(0) \underbrace{+x(t=0) - e^{-i\mu J_x}x(t=\beta)}_{\text{we have to make it zero.}}\Big).
\end{aligned} \tag{2.76}$$

This is the origin of the twisted boundary condition in (2.73).

References

1. E. Witten, Quantization of Chern-Simons gauge theory with complex gauge group. Commun. Math. Phys. **137**, 29–66 (1991)
2. S. Weinberg, The Quantum Theory of Fields, *The Quantum Theory of Fields 3 Volume Hardback Set*, vol. 1 (Cambridge University Press, Cambridge, 1995)
3. J. Zinn-Justin, *Path Integrals in Quantum Mechanics*, Oxford Graduate Texts (OUP, Oxford, 2010)
4. M. Swanson, *Path Integrals and Quantum Processes* (Elsevier Science, 2012)
5. J. Sakurai, J. Napolitano, *Modern Quantum Mechanics* (Addison-Wesley, Boston, 2011)
6. E. Witten, Constraints on supersymmetry breaking. Nucl. Phys. B **202**, 253 (1982)
7. L. Alvarez-Gaume, Supersymmetry and the Atiyah-Singer index theorem. Commun. Math. Phys. **90**, 161 (1983)
8. D. Friedan, P. Windey, Supersymmetric derivation of the Atiyah-Singer index and the chiral anomaly. Nucl. Phys. B **235**, 395 (1984)

Chapter 3
Three-Dimensional Superconformal Index on $M^2 \times S^1_\beta$

In this chapter, we review recent development on the three-dimensional superconformal index (SCI)

$$\mathcal{I}^{M^2}_{\text{Theory}}(x, \alpha_a) = \text{Tr}_{\mathcal{H}(M^2)}\Big((-1)^{\hat{F}} x'^{\{Q,Q^\dagger\}} x^{\hat{H}+\hat{j}_3} \prod_a \alpha_a^{\hat{f}_a}\Big), \tag{3.1}$$

based on supersymmetric localization principle. In Sect. 3.1, we give the physical meaning for the SCI, and represent it in the path integral formalism. In Sect. 3.2, we turn to define supersymmetric actions on $M^2 \times S^1_\beta$ where β corresponds to the inverse temperature. In Sect. 3.3, we explain the supersymmetric localization principle. We will perform the exact calculations in later chapters based on this technique.

3.1 Superconformal Index

First, we consider the physical meaning of the SCI in operator formalism. After that, we turn to the path integral representation of SCI by quoting the results in Chap. 2.

3.1.1 *Operator Formalism Description*

As one can find in [1–3], it is known that the following operators

$$\hat{H} + \hat{j}_3, \quad \hat{f}_a, \; a = 1, \ldots, N_f \tag{3.2}$$

A. Tanaka, *Superconformal Index on* $\text{RP}^2 \times \text{S}^1$ *and 3D Mirror Symmetry*,
Springer Theses, DOI 10.1007/978-981-10-1398-0_3

commute with both of $\hat{Q}$ and $\hat{Q}^\dagger$, therefore, each operator can play a role of $\hat{J}$ in (2.71) and the SCI (3.1) turns to be generalized index and does not depends on x'. It means that states which satisfy

$$\{\hat{Q}, \hat{Q}^\dagger\}|\text{phys}\rangle = 0 \quad \Leftrightarrow \quad \hat{Q}|\text{phys}\rangle = 0 \tag{3.3}$$

called BPS states [4, 5] only contribute to the SCI. Now, we define subspace of the Hilbert space $\mathcal{H}$:

$$\mathcal{H}^{BPS} := \Big\{ |\text{phys}\rangle \in \mathcal{H} \,\Big|\, \hat{Q}|\text{phys}\rangle = 0 \Big\}. \tag{3.4}$$

Then, we can rewrite SCI as follows

$$\mathcal{I}(x, \alpha_a) = \text{Tr}_{\mathcal{H}^{BPS}}\Big((-1)^{\hat{F}} x^{\hat{H}+\hat{j}_3} \prod_a \alpha_a^{\hat{f}_a} \Big). \tag{3.5}$$

For simplicity, we suppose here the index a runs for $a = 1$ only, and omit this index, then SCI reduces to

$$\mathcal{I}(x, \alpha) = \text{Tr}_{\mathcal{H}^{BPS}}\Big((-1)^{\hat{F}} x^{\hat{H}+\hat{j}_3} \alpha^{\hat{f}} \Big). \tag{3.6}$$

$\hat{H} + \hat{j}_3$ and $\hat{f}$ are conserved charges so we can divide $\mathcal{H}^{BPS}$ into more basic ingredients

$$\mathcal{H}^{BPS}_{J,f} := \left\{ |J, f\rangle \in \mathcal{H}^{BPS} \,\middle|\, \begin{array}{l} (\hat{H} + \hat{j}_3)|J, f\rangle = J|J, f\rangle \\ \hat{f}|J, f\rangle = f|J, f\rangle \end{array} \right\}. \tag{3.7}$$

Then, SCI can be represented by each Witten index of (J, f) sector $I_{(J,f)}$:

$$\mathcal{I}(x, \alpha) = \sum_{J,f} x^J \alpha^f \times \underbrace{\text{Tr}_{\mathcal{H}^{BPS}_{J,f}}(-1)^{\hat{F}}}_{I_{(J,f)}}. \tag{3.8}$$

Therefore, once we know the exact form of the $\mathcal{I}(x, \alpha)$, we can extract the number $I_{(J,f)}$ by Taylor-expanding it around $x = \alpha = 0$. Compared with the usual Witten index I which provides us the number of degenerated "vacua" of whole Hilbert space, SCI gives us finer informations of the theory because it provides us Witten indices with fixed J, f, $I_{(J,f)}$. Of course it is expected that the SCI (3.1) goes back to the usual Witten index just by taking $x = \alpha = 1$, so $I = \sum I_{(J,f)}$ should be satisfied formally.

3.1.2 Path Integral Description

In order to convert to the path integral description, it is useful to introduce parameters $\beta_1, \beta_2, \beta, \mu_a$ as follows.

$$x' = e^{-\beta_1}, \quad x = e^{-\beta_2}, \quad \alpha_a = e^{-i\mu_a}, \quad \beta = \beta_1 + \beta_2. \tag{3.9}$$

By utilizing the $\mathcal{N} = 2$ SUSY algebra [1–3, 6], we get the relation

$$\{\hat{Q}, \hat{Q}^\dagger\} = \hat{H} + \hat{R} - \hat{j}_3, \tag{3.10}$$

where $\hat{R}$ is called *R-charge*.[1] Then we can rewrite the SCI as follows:

$$\mathcal{I}(x, \alpha_a) = \text{Tr}_{\mathcal{H}}\Big((-1)^{\hat{F}} e^{-\beta \hat{H}} \cdot e^{-\beta_1(\hat{R}-\hat{j}_3)} e^{-\beta_2 \hat{j}_3} e^{-\sum_a i\mu_a \hat{f}_a}\Big). \tag{3.11}$$

As we have already mentioned in Chap. 2, the $e^{-\beta \hat{H}}$ generates translation along the β circle, $(-1)^{\hat{F}}$ makes all fields periodic, and other insertions $e^{-\beta_1(\hat{R}-\hat{j}_3)} e^{-\beta_2 \hat{j}_3} e^{\sum_a i\mu_a \hat{f}_a}$ define twisted boundary condition for each field: (See also (2.73).)

$$x(t+\beta) = e^{\beta_1(\hat{R}-\hat{j}_3)} e^{\beta_2 \hat{j}_3} e^{\sum_a i\mu_a \hat{f}_a} x(t), \quad \text{for boson,} \tag{3.12}$$

$$\Psi_\pm(t+\beta) = e^{\beta_1(\hat{R}-\hat{j}_3)} e^{\beta_2 \hat{j}_3} e^{\sum_a i\mu_a \hat{f}_a} \Psi_\pm(t), \quad \text{for fermion.} \tag{3.13}$$

Therefore, by repeating the derivation of the path integral description of the Witten index or generalized index, we arrive at the path integral definition:

$$\mathcal{I}(x, \alpha_a) \sim \int \left(\prod_{t \in [0,\beta]} dx(t) d\Psi_+(t) d\Psi_-(t) \right) e^{-S_b - S_f}, \tag{3.14}$$

with conditions (3.12) and (3.13).

To quantum field theory The above explanation is almost correct, but more precisely speaking, we should add two spacial dimensions represented by x^i $(i = 1, 2)$ which is a set of coordinates for certain two-dimensional manifold M^2, and consider not quantum mechanical degrees of freedom but quantum field theoretical degrees of freedom:

$$x(t) \to \phi(x^i, t), \qquad \Psi_+(t) \to \overline{\psi}(x^i, t), \quad \Psi_-(t) \to \psi(x^i, t). \tag{3.15}$$

And of course the twisted boundary conditions (3.12) and (3.13) are lifted to

[1] We will assign R-charge to each field later. (See Table 3.1.)

Table 3.1 Charge assignments for each field. $\hat{R}$ is the R-charge appeared in (3.10)

Killing spinor	ϵ	$\overline{\epsilon}$									
Spin	1/2	1/2									
$\hat{R}$	+1	−1									
Field	A_μ	σ	λ	$\overline{\lambda}$	D	ϕ	$\overline{\phi}$	ψ	$\overline{\psi}$	F	$\overline{F}$
Spin	1	0	1/2	1/2	0	0	0	1/2	1/2	0	0
$\hat{R}$	0	0	+1	−1	0	$-\Delta$	Δ	$-(\Delta-1)$	$\Delta-1$	$-(\Delta-2)$	$\Delta-2$

$$\phi(x^i, t+\beta) = e^{\beta_1(\hat{R}-\hat{j}_3)} e^{\beta_2 \hat{j}_3} e^{\sum_a i\mu_a \hat{f}_a} \phi(x^i, t), \quad \text{for bosons,} \tag{3.16}$$

$$\psi(x^i, t+\beta) = e^{\beta_1(\hat{R}-\hat{j}_3)} e^{\beta_2 \hat{j}_3} e^{\sum_a i\mu_a \hat{f}_a} \psi(x^i, t), \quad \text{for fermions.} \tag{3.17}$$

Finally, we get the path integral representation of SCI (3.1) as

$$\mathcal{I}(x, \alpha_a) = \int_{\text{b.c. (3.16) and (3.17)}} \mathcal{D}\phi \mathcal{D}\psi \mathcal{D}\overline{\psi}\, e^{-S_b - S_f}. \tag{3.18}$$

3.2 Supersymmetric Field Theories on Curved Manifold $M^2 \times S^1_\beta$

Let us begin to discuss the main part of this thesis. Our main interest is to calculate SCI (3.1) by using the path integral (3.18) under the twisted boundary conditions (3.16) and (3.17). To do so, it is useful to make the supersymmetry with the off-shell formalism. We use so-called three-dimensional $\mathcal{N} = 2$ supersymmetries. There are two irreducible representations, called vector multiplet and matter multiplet. From now on, we take two-dimensional manifold M^2 as round sphere S^2 or real projective space RP^2:

$$S^2 : ds^2_{M^2} = d\vartheta^2 + \sin^2 \vartheta d\varphi^2, \quad \begin{cases} 0 \leq \vartheta \leq \pi \\ 0 \leq \varphi < 2\pi \end{cases}, \tag{3.19}$$

$$RP^2 : ds^2_{M^2} = d\vartheta^2 + \sin^2 \vartheta d\varphi^2, \quad \begin{cases} 0 \leq \vartheta \leq \pi \\ 0 \leq \varphi < 2\pi \\ (\vartheta, \varphi) \sim (\pi - \vartheta, \pi + \varphi) \end{cases}. \tag{3.20}$$

As one can see, the difference between S^2 and RP^2 is just the global information coming from antipodal identification $(\vartheta, \varphi) \sim (\pi - \vartheta, \pi + \varphi)$. Therefore, once we can construct a supersymmetry on S^2, if its representation is based on local Lagrangian description, we can project it into the theory on RP^2. The projection might looks trivial, however the life is not so simple. For example, in mathematical point of view, we have the following 2nd homology groups

$$H_2(S^2) = Z, \quad H_2(RP^2) = 0. \tag{3.21}$$

This means that the classical gauge field on S^2 is labeled by the 1st Chern number, or equivalently monopole number. In addition to it, the fundamental groups are as follows.

$$\pi_1(S^2) = 0, \quad \pi_1(RP^2) = Z_2. \tag{3.22}$$

This fact means that the classical gauge field on RP^2 is labeled by the Z_2-holonomy, or equivalently (discretized) Wilson line phases.[2]

3.2.1 *Our Convention for Spinors*

We consider the following dreibein:

$$e^1 = d\vartheta, \quad e^2 = \sin\vartheta d\varphi, \quad e^3 = dy. \tag{3.23}$$

We use alphabetical indices $a, b, c, \dots$ for the local Lorentz indices.

Covariant derivative The three-dimensional covariant derivative is defined by

$$\nabla_\mu = \partial_\mu + \frac{1}{4}\omega_\mu^{ab}\hat{\mathcal{J}}_{ab} \tag{3.24}$$

where ω_μ^{ab} is the spin connection computed from the dreibein (3.23),

$$de^a + \omega^{ab}\wedge e^b = 0, \quad \omega^{ba} = -\omega^{ab}, \quad \omega^{ab} = \omega_\mu^{ab}dx^\mu. \tag{3.25}$$

$\hat{\mathcal{J}}_{ab}$ are Lorentz generators of the fields characterized by its spin:

$$\begin{array}{ll} \text{spin } 0 & \Rightarrow \hat{\mathcal{J}}_{ab} = 0, \\ \text{spin } 1/2 & \Rightarrow \hat{\mathcal{J}}_{ab} = \gamma_{ab}, \\ \text{spin } 1 & \Rightarrow (\hat{\mathcal{J}}_{ab})^c{}_d = 2(\delta^{ac}\delta^{bd} - \delta^{bc}\delta^{ad}), \end{array} \tag{3.26}$$

where γ_{ab} are antisymmetrized gamma matrices defined in (3.27).

Gamma matrices The gamma matrices γ_a are defined by the Pauli matrices

$$\gamma_1 = \begin{pmatrix} 0 & 1 \\ 1 & 0 \end{pmatrix}, \quad \gamma_2 = \begin{pmatrix} 0 & -i \\ i & 0 \end{pmatrix}, \quad \gamma_3 = \begin{pmatrix} 1 & 0 \\ 0 & -1 \end{pmatrix}, \quad \gamma_{ab} = \frac{1}{2}(\gamma_a\gamma_b - \gamma_b\gamma_a). \tag{3.27}$$

Spinor bilinear Our convention is as follows. Let us denote generic spinors by $\epsilon, \bar{\epsilon}$, and λ. We take spinor bilinears as

$$\epsilon\lambda = \begin{pmatrix} \epsilon_1 & \epsilon_2 \end{pmatrix} \begin{pmatrix} 0 & 1 \\ -1 & 0 \end{pmatrix} \begin{pmatrix} \lambda_1 \\ \lambda_2 \end{pmatrix}, \quad \epsilon\gamma_a\lambda = \begin{pmatrix} \epsilon_1 & \epsilon_2 \end{pmatrix} \begin{pmatrix} 0 & 1 \\ -1 & 0 \end{pmatrix} \gamma_a \begin{pmatrix} \lambda_1 \\ \lambda_2 \end{pmatrix}.$$

Using this convention, one can prove the following formulas:

[2]However, there is another possibility. See [7] for example. We will comment on it in the final chapter.

$$\epsilon\lambda = (-1)^{1+|\epsilon|\cdot|\lambda|}\lambda\epsilon, \quad \epsilon\gamma_a\lambda = (-1)^{|\epsilon|\cdot|\lambda|}\lambda\gamma_a\epsilon, \quad (\gamma_a\epsilon)\lambda = -\epsilon\gamma_a\lambda,$$
$$\overline{\epsilon}(\epsilon\lambda) + (-1)^{1+|\epsilon|\cdot|\overline{\epsilon}|}\epsilon(\overline{\epsilon}\lambda) + (\overline{\epsilon}\epsilon)\lambda = 0,$$
$$(-1)^{1+|\epsilon|\cdot|\overline{\epsilon}|}\epsilon(\overline{\epsilon}\lambda) + 2(\overline{\epsilon}\epsilon)\lambda + (-1)^{1+|\lambda|\cdot|\epsilon|}(\overline{\epsilon}\gamma_a\lambda)\gamma^a\epsilon = 0,$$

where $|\epsilon|$ means the spinor ϵ's statistics such that $|\epsilon| = 0$ for a bosonic ϵ and $|\epsilon| = 1$ for a fermonic ϵ.

3.2.2 *Killing Spinors*

Now what we want to do is to construct SUSY QFTs on $M^2 \times S^1_\beta$ with the metric

$$ds^2 = ds^2_{M^2} + dt^2. \tag{3.28}$$

As well known, so-called superspace formalism is very useful to construct SUSY theories on flat space [8]. However, the curved superspace formalism is still under construction. (See [6, 9] for theories on two or three spheres.) So we take an ad-hoc way here.[3] In order to construct supersymmetry, it is necessary to construct so-called Killing spinors [10]. With our metric (3.28) and dreibein (3.23), the following two spinors

$$\epsilon(\vartheta, \varphi, y) = e^{\frac{1}{2}(y+i\varphi)}\begin{pmatrix}\cos\frac{\vartheta}{2}\\ \sin\frac{\vartheta}{2}\end{pmatrix}, \quad \overline{\epsilon}(\vartheta, \varphi, y) = e^{\frac{-1}{2}(y+i\varphi)}\begin{pmatrix}\sin\frac{\vartheta}{2}\\ \cos\frac{\vartheta}{2}\end{pmatrix} \tag{3.29}$$

satisfy the following equations

$$\nabla_\mu\epsilon = \frac{1}{2}\gamma_\mu\gamma_3\epsilon, \quad \nabla_\mu\overline{\epsilon} = \frac{-1}{2}\gamma_\mu\gamma_3\overline{\epsilon}. \tag{3.30}$$

These spinors are Killing spinors in our case. In later discussion, we use these spinors $\overline{\epsilon}$, ϵ as parameters of supersymmetry.

3.2.3 $\mathcal{N} = 2$ *Vector Multiplet*

Vector multiplet is constructed from a gauge field A_μ, an adjoint scalar field σ, an auxiliary field D, and adjoint 2-component spinors $\overline{\lambda}$, λ:

$$V := (A_\mu, \sigma, D \mid \overline{\lambda}, \lambda). \tag{3.31}$$

[3]There is a systematical way via supergravity theory [10]. We will not consider it here, but the result is equivalent.

$\mathcal{N} = 2$ supersymmetry is defined as follows [11]:

$$\delta_\epsilon A_\mu = -\frac{i}{2}\overline{\lambda}\gamma_\mu\epsilon, \quad \delta_{\overline{\epsilon}} A_\mu = -\frac{i}{2}\overline{\epsilon}\gamma_\mu\lambda, \tag{3.32}$$

$$\delta_\epsilon\sigma = +\frac{1}{2}\overline{\lambda}\epsilon, \quad \delta_{\overline{\epsilon}}\sigma = +\frac{1}{2}\overline{\epsilon}\lambda, \tag{3.33}$$

$$\delta_\epsilon\lambda = \frac{1}{2}\gamma^{\mu\nu}\epsilon F_{\mu\nu} - D\epsilon + i\gamma^\mu\epsilon\mathcal{D}_\mu\sigma + \frac{2i}{3}\sigma\gamma^\mu\nabla_\mu\epsilon, \quad \delta_{\overline{\epsilon}}\lambda = 0, \tag{3.34}$$

$$\delta_\epsilon\overline{\lambda} = 0, \quad \delta_{\overline{\epsilon}}\overline{\lambda} = \frac{1}{2}\gamma^{\mu\nu}\overline{\epsilon}F_{\mu\nu} + D\overline{\epsilon} - i\gamma^\mu\overline{\epsilon}\mathcal{D}_\mu\sigma - \frac{2i}{3}\sigma\gamma^\mu\nabla_\mu\overline{\epsilon}, \tag{3.35}$$

$$\delta_\epsilon D = +\frac{i}{2}\mathcal{D}_\mu\overline{\lambda}\gamma^\mu\epsilon - \frac{i}{2}[\overline{\lambda}\epsilon, \sigma] + \frac{i}{6}\overline{\lambda}\gamma^\mu\nabla_\mu\epsilon, \delta_{\overline{\epsilon}} D = -\frac{i}{2}\overline{\epsilon}\gamma^\mu\mathcal{D}_\mu\lambda + \frac{i}{2}[\overline{\epsilon}\lambda, \sigma] - \frac{i}{6}\nabla_\mu\overline{\epsilon}\gamma^\mu\lambda. \tag{3.36}$$

The covariant derivative $\mathcal{D}_\mu$ is defined as

$$\mathcal{D}_\mu = \nabla_\mu - i[A_\mu, \circ]. \tag{3.37}$$

One can verify the following algebraic structure:

$$\{\delta_\epsilon, \delta_\epsilon\} = 0, \quad \{\delta_{\overline{\epsilon}}, \delta_{\overline{\epsilon}}\} = 0, \tag{3.38}$$

$$\{\delta_\epsilon, \delta_{\overline{\epsilon}}\}A_\mu = \xi^\nu\partial_\nu A_\mu + \partial_\mu\xi^\nu A_\nu + \mathcal{D}_\mu\Lambda, \tag{3.39}$$

$$\{\delta_\epsilon, \delta_{\overline{\epsilon}}\}\sigma = \xi^\mu\partial_\mu\sigma + i[\Lambda, \sigma], \tag{3.40}$$

$$\{\delta_\epsilon, \delta_{\overline{\epsilon}}\}\lambda = \xi^\mu\partial_\mu\lambda + \frac{1}{4}\Theta_{\mu\nu}\gamma^{\mu\nu}\lambda + i[\Lambda, \lambda] + \alpha\lambda, \tag{3.41}$$

$$\{\delta_\epsilon, \delta_{\overline{\epsilon}}\}\overline{\lambda} = \xi^\mu\partial_\mu\overline{\lambda} + \frac{1}{4}\Theta_{\mu\nu}\gamma^{\mu\nu}\overline{\lambda} + i[\Lambda, \overline{\lambda}] - \alpha\overline{\lambda}, \tag{3.42}$$

$$\{\delta_\epsilon, \delta_{\overline{\epsilon}}\}D = \xi^\mu\partial_\mu D + i[\Lambda, D]. \tag{3.43}$$

Equations (3.39)–(3.43) relations mean

$$\{\delta_\epsilon, \delta_{\overline{\epsilon}}\} = \delta^\xi_{\text{Translation}} + \delta^\Theta_{\text{Rotation}} + \delta^\Lambda_{\text{Gauge transformation}} + \delta^\alpha_{\text{R-symmetry}}, \tag{3.44}$$

where each parameter is defined as follows.

$$\xi^\mu = i\overline{\epsilon}\gamma^\mu\epsilon, \tag{3.45}$$

$$\Theta^{\mu\nu} = \nabla^{[\mu}\xi^{\nu]} + \xi^\lambda\omega_\lambda^{\mu\nu}, \tag{3.46}$$

$$\Lambda = -A_\mu\xi^\mu + \sigma\overline{\epsilon}\epsilon, \tag{3.47}$$

$$\alpha = \frac{i}{3}(\nabla_\mu\overline{\epsilon}\gamma^\mu\epsilon - \overline{\epsilon}\gamma^\mu\nabla_\mu\epsilon). \tag{3.48}$$

3.2.4 $\mathcal{N} = 2$ Matter Multiplet

Matter multiplet is constructed from scalar fields $\phi, \overline{\phi}$, spinor fields $\psi, \overline{\psi}$, and auxiliary fields $F, \overline{F}$:

$$\Phi := (\phi, F \mid \psi), \quad \overline{\Phi} := (\overline{\phi}, \overline{F} \mid \overline{\psi}). \tag{3.49}$$

We can couple these fields to the vector multiplet (3.31) in supersymmetric way. In addition to it, we can assign arbitrary conformal dimension, or equivalently R-charge Δ to the matter multiplet (3.49). $\mathcal{N} = 2$ supersymmetry is defined as follows [11]:

$$\delta_\epsilon \phi = 0, \quad \delta_{\overline{\epsilon}} \phi = \overline{\epsilon} \psi, \tag{3.50}$$
$$\delta_\epsilon \overline{\phi} = \epsilon \overline{\psi}, \quad \delta_{\overline{\epsilon}} \overline{\phi} = 0, \tag{3.51}$$
$$\delta_\epsilon \psi = i \gamma^\mu \epsilon \mathcal{D}^A_\mu \phi + i \epsilon \sigma \phi + \frac{2\Delta i}{3} \gamma^\mu \nabla_\mu \epsilon \, \phi, \quad \delta_{\overline{\epsilon}} \psi = \overline{\epsilon} F, \tag{3.52}$$
$$\delta_\epsilon \overline{\psi} = \overline{F} \epsilon, \quad \delta_{\overline{\epsilon}} \overline{\psi} = i \gamma^\mu \overline{\epsilon} \mathcal{D}^A_\mu \overline{\phi} + i \overline{\phi} \sigma \overline{\epsilon} + \frac{2\Delta i}{3} \overline{\phi} \gamma^\mu \nabla_\mu \overline{\epsilon}, \tag{3.53}$$
$$\delta_\epsilon F = \epsilon (i \gamma^\mu \mathcal{D}^A_\mu \psi - i \sigma \psi - i \lambda \phi) + \frac{i}{3} (2\Delta - 1) \nabla_\mu \epsilon \gamma^\mu \psi, \quad \delta_{\overline{\epsilon}} F = 0, \tag{3.54}$$
$$\delta_\epsilon \overline{F} = 0, \quad \delta_{\overline{\epsilon}} \overline{F} = \overline{\epsilon} (i \gamma^\mu \mathcal{D}^A_\mu \overline{\psi} - i \overline{\psi} \sigma + i \overline{\phi \lambda}) + \frac{i}{3} (2\Delta - 1) \nabla_\mu \overline{\epsilon} \gamma^\mu \overline{\psi}. \tag{3.55}$$

We define the covariant derivative $\mathcal{D}^A_\mu$ as

$$\mathcal{D}^A_\mu \Phi = \mathcal{D}_\mu \Phi - i A_\mu \Phi, \quad \mathcal{D}^A_\mu \overline{\Phi} = \mathcal{D}_\mu \overline{\Phi} + i \overline{\Phi} A_\mu. \tag{3.56}$$

One can verify the following relations:

$$\{\delta_\epsilon, \delta_\epsilon\} = 0, \quad \{\delta_{\overline{\epsilon}}, \delta_{\overline{\epsilon}}\} = 0, \tag{3.57}$$
$$\{\delta_{\overline{\epsilon}}, \delta_\epsilon\} \phi = \xi^\mu \partial_\mu \phi + i \Lambda \phi - \Delta \alpha \phi, \tag{3.58}$$
$$\{\delta_{\overline{\epsilon}}, \delta_\epsilon\} \overline{\phi} = \xi^\mu \partial_\mu \overline{\phi} - i \overline{\phi} \Lambda + \Delta \alpha \overline{\phi}, \tag{3.59}$$
$$\{\delta_{\overline{\epsilon}}, \delta_\epsilon\} \psi = \xi^\mu \partial_\mu \psi + \frac{1}{4} \Theta_{\mu\nu} \gamma^{\mu\nu} \psi + i \Lambda \psi + (1 - \Delta) \alpha \psi, \tag{3.60}$$
$$\{\delta_{\overline{\epsilon}}, \delta_\epsilon\} \overline{\psi} = \xi^\mu \partial_\mu \overline{\psi} + \frac{1}{4} \Theta_{\mu\nu} \gamma^{\mu\nu} \overline{\psi} - i \overline{\psi} \Lambda + (\Delta - 1) \alpha \overline{\psi}, \tag{3.61}$$
$$\{\delta_{\overline{\epsilon}}, \delta_\epsilon\} F = \xi^\mu \partial_\mu F + i \Lambda F + (2 - \Delta) \alpha F, \tag{3.62}$$
$$\{\delta_{\overline{\epsilon}}, \delta_\epsilon\} \overline{F} = \xi^\mu \partial_\mu \overline{F} - i \overline{F} \Lambda + (\Delta - 2) \alpha \overline{F}. \tag{3.63}$$

Of course, we can interpret these relations as the one in (3.44).

3.2.5 SUSY Invariant Lagrangians

We summarize here the SUSY invariant Lagrangians which will be important in later discussion of this thesis.

Supersymmetric Yang–Mills term This action is automatically SUSY invariant because it can be rewrite as $S_{YM} = \delta_\epsilon \mathcal{V}_V = \delta_{\overline{\epsilon}} \tilde{\mathcal{V}}_V$ for certain $\mathcal{V}_V$, $\tilde{\mathcal{V}}_V$ [12], and thanks to the nilpotent natures of δ_ϵ, $\delta_{\overline{\epsilon}}$ (3.38).

$$S_{YM} = \int d^3x \sqrt{g}\, \mathrm{Tr}\Big(+ \frac{1}{2} F_{\mu\nu} F^{\mu\nu} + D^2 + \mathcal{D}_\mu \sigma \cdot \mathcal{D}^\mu \sigma + \epsilon^{3\rho\sigma} \sigma F_{\rho\sigma} + \sigma^2 + i\overline{\lambda}\gamma^\mu \mathcal{D}_\mu \lambda - i\overline{\lambda}[\lambda, \sigma] - \frac{i}{2} \overline{\lambda}\gamma_3 \lambda \Big) \tag{3.64}$$

Supersymmatric matter kinetic term This action is automatically SUSY invariant because of the fact that it can be rewrite as $S_{mat} = \delta_\epsilon \mathcal{V}_M = \delta_{\overline{\epsilon}} \tilde{\mathcal{V}}_M$ for certain $\mathcal{V}_M$, $\tilde{\mathcal{V}}_M$ [12], and thanks to the nilpotent natures of δ_ϵ, $\delta_{\overline{\epsilon}}$ (3.57).

$$S_{mat} = \int d^3x \sqrt{g} \Big(- i(\overline{\psi}\gamma^\mu \mathcal{D}^A_\mu \psi) + i(\overline{\psi}\sigma\psi) - i\overline{\phi}(\overline{\lambda}\psi) - \frac{i(2\Delta - 1)}{2} (\overline{\psi}\gamma_3\psi) + \overline{F}F + i(\overline{\psi}\lambda)\phi + \mathcal{D}^A_\mu \overline{\phi} \mathcal{D}^\mu_A \phi + \overline{\phi}\sigma^2\phi + i\overline{\phi} D\phi - (2\Delta - 1)\overline{\phi}\mathcal{D}^A_3 \phi - \frac{\Delta(2\Delta - 1)}{2} \overline{\phi}\phi + \frac{\Delta}{4} R\overline{\phi}\phi \Big) \tag{3.65}$$

Superpotential term In order to construct this term systematically, for example, superspace formalism [6, 9] is useful. We will use such term in later discussion, however the result does not depends on this term thanks to the powerful calculation method, localization. So we do not comment on them here.

3.3 Supersymmetric Localization Techniques

The mirror symmetry conjecture predicts an equivalence between two theories with *non-trivial interactions*. Therefore, the *exact check* sounds difficult in usual sense. However, a very powerful method had been introduced in [13] which provides an exact calculation method for path integrals of interacting SUSY theories on flat 4d space. This method is called *supersymmetric localization technique*. After the discovery of it, this technique had been extended to the SUSY theories on four-sphere [14], three-sphere [11, 15, 16], and deformed spheres [17–19], and other various dimensional manifolds. We utilize this method on $M^2 \times S^1_\beta$ [3, 12, 20] which give SCI. M^2 represents two-sphere S^2 or real projective plane RP^2. The lower index β corresponds to the inverse temperature. The localization technique is applicable if there are

$$\text{A SUSY: } \delta, \quad \text{A functional: } \mathcal{V},$$

$$\text{A SUSY exact action: } S = \delta\mathcal{V}, \quad \text{such that } \begin{cases} \delta S = 0 \\ S_{boson} \geq 0 \end{cases} .$$

Note that the actions defined in (3.64) and (3.65) satisfy this condition. Then, the path integral

$$\int \mathcal{D}\phi\mathcal{D}\psi\, e^{-S[\phi,\psi]} \tag{3.66}$$

can be computed from

$$I(t) = \int \mathcal{D}\phi\mathcal{D}\psi\, e^{-tS[\phi,\psi]} \tag{3.67}$$

because $I(t)$ does not depend on t. One can derive this fact as follows.

$$\begin{aligned} \frac{dI(t)}{dt} &= \int \mathcal{D}\phi\mathcal{D}\psi(-S)\, e^{-tS} \\ &- \int \mathcal{D}\phi\mathcal{D}\psi(-\delta\mathcal{V})\, e^{-tS} \\ &= -\int \mathcal{D}\phi\mathcal{D}\psi\, \delta(\mathcal{V}e^{-tS}) = 0. \end{aligned} \tag{3.68}$$

In order to perform the path integral (3.67), we can take the ultimate limit $t \to \infty$ because $I(t)$ does not depend on t! Then, the field configurations ϕ_0, ψ_0 which give

$$S[\phi_0, \psi_0] = \frac{\partial S}{\partial\phi}[\phi_0, \psi_0] = \frac{\partial S}{\partial\psi}[\phi_0, \psi_0] = 0, \tag{3.69}$$

dominate. We call them *locus*. We can expand each field around the locus:

$$\phi = \phi_0 + \frac{1}{\sqrt{t}}\tilde{\phi}, \quad \psi = \psi_0 + \frac{1}{\sqrt{t}}\tilde{\psi}, \tag{3.70}$$

then the action becomes

$$tS[\phi, \psi] = \underbrace{\frac{1}{2}\tilde{\phi}\frac{\partial S}{\partial\phi\partial\phi}[\phi_0, \psi_0]\tilde{\phi} + \frac{1}{2}\tilde{\psi}\frac{\partial S}{\partial\psi\partial\psi}[\phi_0, \psi_0]\tilde{\psi}}_{:=\tilde{S}[\phi_0,\psi_0;\tilde{\phi},\tilde{\psi}]} + \mathcal{O}(t^{-1/2}). \tag{3.71}$$

By taking $t \to \infty$, only the first two parts contribute. We define it as $\tilde{S}[\psi_0, \psi_0, \tilde{\psi}, \tilde{\psi}]$. After taking into account the cancellation of t in the measure $\mathcal{D}\phi\mathcal{D}\psi$, the original path integral can be calculated by summing up all Gaussian contributions around the locus.

$$\int \mathcal{D}\phi\mathcal{D}\psi\, e^{-S[\phi,\psi]} = \sum_{\phi_0,\psi_0} \int \mathcal{D}\tilde{\phi}\mathcal{D}\tilde{\psi}\, e^{-\tilde{S}[\phi_0,\psi_0;\tilde{\phi},\tilde{\psi}]}. \tag{3.72}$$

Roughly speaking, this is the analog of the steepest decent method in usual integral on complex plane. We will utilize this method, and perform the exact check of (1.3).

References

1. J. Bhattacharya, S. Bhattacharyya, S. Minwalla, S. Raju, Indices for superconformal field theories in 3, 5 and 6 Dimensions. J. High Energy Phys. **0802**, 064 (2008). arXiv:0801.1435 [hep-th]
2. J. Bhattacharya, S. Minwalla, Superconformal indices for N = 6 Chern Simons theories. J. High Energy Phys. **0901**, 014 (2009). arXiv:0806.3251 [hep-th]
3. S. Kim, The complete superconformal index for N = 6 Chern-Simons theory. Nucl. Phys. B **821**, 241–284 (2009). arXiv:0903.4172 [hep-th]
4. E. Bogomolny, Stability of classical solutions. Sov. J. Nucl. Phys. **24**, 449 (1976)
5. M. Prasad, C.M. Sommerfield, An exact classical solution for the 't Hooft monopole and the Julia-Zee dyon. Phys. Rev. Lett. **35**, 760–762 (1975)
6. I. Samsonov, D. Sorokin, Gauge and matter superfield theories on S^2. J. High Energy Phys. **1409**, 097 (2014). arXiv:1407.6270 [hep-th]
7. A. Tanaka, H. Mori, T. Morita, Abelian 3d mirror symmetry on $\mathrm{RP}^2 \times \mathrm{S}^1$ with $N_f = 1$. J. High Energy Phys. **09**, 154 (2015). arXiv:1505.07539 [hep-th]
8. J. Wess, J. Bagger, Supersymmetry and supergravity (1992)
9. I. Samsonov, D. Sorokin, Superfield theories on S^3 and their localization. J. High Energy Phys. **1404**, 102 (2014). arXiv:1401.7952 [hep-th]
10. C. Closset, T.T. Dumitrescu, G. Festuccia, Z. Komargodski, Supersymmetric field theories on three-manifolds. J. High Energy Phys. **1305**, 017 (2013). arXiv:1212.3388 [hep-th]
11. N. Hama, K. Hosomichi, S. Lee, Notes on SUSY gauge theories on three-sphere. J. High Energy Phys. **1103**, 127 (2011). arXiv:1012.3512 [hep-th]
12. A. Tanaka, H. Mori, T. Morita, Superconformal index on $RP^2 \times S^1$ and mirror symmetry. arXiv:1408.3371 [hep-th]
13. E. Witten, Topological quantum field theory. Commun. Math. Phys. **117**, 353 (1988)
14. V. Pestun, Localization of gauge theory on a four-sphere and supersymmetric Wilson loops. Commun. Math. Phys. **313**, 71–129 (2012). arXiv:0712.2824 [hep-th]
15. A. Kapustin, B. Willett, I. Yaakov, Exact results for Wilson loops in superconformal Chern-Simons theories with matter. J. High Energy Phys. **1003**, 089 (2010). arXiv:0909.4559 [hep-th]
16. D.L. Jafferis, The exact superconformal R-symmetry extremizes Z. J. High Energy Phys. **1205**, 159 (2012). arXiv:1012.3210 [hep-th]
17. Y. Imamura, D. Yokoyama, N = 2 supersymmetric theories on squashed three-sphere. Phys. Rev. D **85**, 025015 (2012). arXiv:1109.4734 [hep-th]
18. N. Hama, K. Hosomichi, S. Lee, SUSY gauge theories on squashed three-spheres. J. High Energy Phys. **1105**, 014 (2011). arXiv:1102.4716 [hep-th]
19. A. Tanaka, Localization on round sphere revisited. J. High Energy Phys. **1311**, 103 (2013). arXiv:1309.4992 [hep-th]
20. Y. Imamura, S. Yokoyama, Index for three dimensional superconformal field theories with general R-charge assignments. J. High Energy Phys. **1104**, 007 (2011). arXiv:1101.0557 [hep-th]

Chapter 4
Localization Calculous of SCI with $M^2 = S^2$

In this chapter, we mainly review the calculations performed in [1–4]. If we consider the $U(1)$ gauge theory, the action (3.64) itself defines free theory. It may sound not so interesting, however we can turn on the gauge coupling in matter action (3.65) like usual QED, this is nontrivial theory. Once we consider non-abelian gauge group, there are some different points in the argument, however the essence is same. Therefore, we focus on the gauge theory with abelian gauge field from now on.

4.1 Vector Multiplet

Locus Now, let us remind that the Lagrangian (3.64), SUSY exact Lagrangian for vector multiplet. One can easily check that the Lagrangian can be deformed to

$$\mathcal{L}_{\text{YM}} = \mathcal{F}_\mu \mathcal{F}^\mu + D^2 + i\overline{\lambda}\gamma^\mu \mathcal{D}_\mu \lambda - \frac{i}{2}\overline{\lambda}\gamma_3\lambda,$$
$$\mathcal{F}^\mu = \frac{1}{2}\epsilon^{\mu\rho\sigma} F_{\rho\sigma} + \partial^\mu \sigma + \delta_3^\mu \sigma. \tag{4.1}$$

The bosonic terms are obviously positive definite. Therefore, we can use this action as the $S = \delta\mathcal{V}$ term in (3.72), and the localization locus, which corresponds to the pair of configurations ϕ_0, ψ_0 in (3.69), is determined by the following equations:

$$0 = \mathcal{F}^\mu = D, \quad \lambda = \overline{\lambda} = 0. \tag{4.2}$$

We can solve this equation by taking

$$A = A_{\text{mon}} + \frac{\theta}{\beta} dt, \quad \sigma = -\frac{B}{2}, \tag{4.3}$$

A. Tanaka, *Superconformal Index on* $\text{RP}^2 \times \text{S}^1$ *and 3D Mirror Symmetry*, Springer Theses, DOI 10.1007/978-981-10-1398-0_4

where A_{mon} is defined as

$$A_{\text{mon}} = \frac{B}{2}(\kappa - \cos\vartheta)d\varphi, \quad \kappa = \begin{cases} +1 \text{ for } 0 \leq \vartheta < \pi \\ -1 \text{ for } 0 < \vartheta \leq \pi \end{cases}. \tag{4.4}$$

Thanks to the gauge symmetry, the parameters B, θ are constrained as[1]

$$B \in Z, \quad \theta \in [0, 2\pi]. \tag{4.5}$$

As explained in the previous chapter, in the context of the supersymmetric localization, we expand field V around the locus V_0 which is parametrized by B, θ:

$$V = V_0[B,\theta] + \frac{1}{\sqrt{t}}\tilde{V}, \quad \text{where} \quad V_0[B,\theta] = \left(A_{\text{mon}} + \frac{\theta}{\beta}dt, -\frac{B}{2}, 0\middle|0, 0\right) \tag{4.6}$$

and $\tilde{V}$ represents fluctuation. The original path integral should be organized by the summation over $B \in Z$, integral over $\theta \in [0, 2\pi]$ and path integral over the fluctuation $\tilde{V}$:

$$\int \mathcal{D}V e^{-S_{YM}[V]} = \sum_{B\in Z} \int_0^{2\pi} \frac{d\theta}{2\pi} \int \mathcal{D}\tilde{V} e^{-\tilde{S}_{YM}[\tilde{V}]}. \tag{4.7}$$

Action for the fluctuation $\tilde{V}$ We show here the action $\tilde{S}_{YM}[\tilde{V}]$ explicitly.

$$\tilde{S}_{\text{boson}} = \int dt \int \sin\vartheta d\vartheta d\varphi \Big(\frac{1}{2}[\partial_\mu \tilde{A}_\nu - \partial_\nu \tilde{A}_\mu]^2 + (\partial_\mu \tilde{\sigma})^2 + \epsilon^{3\mu\nu}\tilde{\sigma}[\partial_\mu \tilde{A}_\nu - \partial_\nu \tilde{A}_\mu] + \tilde{\sigma}^2\Big), \tag{4.8}$$

$$\tilde{S}_{\text{fermion}} = \int dt \int \sin\vartheta d\vartheta d\varphi \Big(i\tilde{\bar{\lambda}}\gamma^\mu \nabla_\mu \tilde{\lambda} - \frac{i}{2}\tilde{\bar{\lambda}}\gamma_3\tilde{\lambda}\Big). \tag{4.9}$$

For later use, we will omit $\tilde{}$ from now on, and decompose the three-dimensional gauge field A_μ to the S^1_β component A_y and 1-form on S^2 $A_2 = A_\vartheta d\vartheta + A_\varphi d\varphi$. Then, the bosonic Lagrangian reduces to

$$\int dy \int \begin{pmatrix} A_2 \\ A_t \\ \sigma \end{pmatrix}^T \wedge *_2 \begin{pmatrix} - *_2 d_2 *_2 d_2 - \partial_y^2 & \partial_y d_2 & - *_2 d_2 \\ \partial_y *_2 d_2 *_2 & - *_2 d_2 *_2 d_2 & 0 \\ *_2 d_2 & 0 & - *_2 d_2 *_2 d_2 - \partial_y^2 + 1 \end{pmatrix} \begin{pmatrix} A_2 \\ A_t \\ \sigma \end{pmatrix}, \tag{4.10}$$

where $*_2$ is the Hodge star operator [5–7] on S^2 defined by

[1]The reason for $B \in Z$ is explained in the Appendix B. The condition for the θ can be also derived by the gauge symmetry.

$$*_2 1 = \sin\vartheta d\vartheta \wedge d\varphi, \quad *_2 d\vartheta = \sin\vartheta d\varphi, \quad *_2 d\varphi = -d\vartheta, \quad *_2 \sin\vartheta d\vartheta \wedge d\varphi = 1, \tag{4.11}$$

and d_2 is the exterior derivative along S^2:

$$d_2 = \frac{\partial}{\partial\vartheta} d\vartheta + \frac{\partial}{\partial\varphi} d\varphi. \tag{4.12}$$

Gauge fixing procedure In order to calculate the path integral, even it is Gaussian, gauge fixing procedure is necessary. In usual procedure, one introduces Fadeev–Popov ghost fields, and construct BRST symmetry, etc. Here, we take simpler root performed in [8–10]. The gauge orbit parametrized by a function η can be represented by

$$\text{Gauge mode:} \begin{pmatrix} A_2^{(\eta)} \\ A_y^{(\eta)} \end{pmatrix} := \begin{pmatrix} i d_2 \eta \\ i \partial_y \eta \end{pmatrix}. \tag{4.13}$$

It gives *zero modes* for the fluctuation integral. We have to get rid of the mode from the path integral. It can be achieved by inserting

$$\delta(A^{(\eta)}). \tag{4.14}$$

into the path integral. However, the precise insertion is

$$\delta(\eta) \tag{4.15}$$

where η is the generator of the gauge transformation mode in (4.13). The Fadeev–Popov determinant is the factor recovering the discrepancy between (4.14) and (4.15):

$$\delta(\eta) = \Delta_{FP}\, \delta(A^{(\eta)}). \tag{4.16}$$

The easiest way to calculate Δ_{FP} is as follows.

$$\begin{aligned} 1 = \int \mathcal{D}A^{(\eta)}\, e^{-\frac{1}{2}\langle A^{(\eta)}, A^{(\eta)}\rangle} &= \Delta_{FP} \int \mathcal{D}\eta\, e^{-\frac{1}{2}\langle A^{(\eta)}, A^{(\eta)}\rangle} \\ &= \Delta_{FP} \int \mathcal{D}\eta\, e^{-\frac{1}{2}\langle d\eta, d\eta\rangle} \\ &= \Delta_{FP} \int \mathcal{D}\eta\, e^{-\frac{1}{2}\langle \eta, d^\dagger d\eta\rangle} \end{aligned} \tag{4.17}$$

where the inner product for the gauge fields is defined by

$$\langle A, B\rangle = \int dy \int \sin\vartheta d\vartheta d\varphi\, A^\mu B_\mu. \tag{4.18}$$

Now, the precise measure for the gauge theory is

$$\Delta_{FP}\, \delta(A^{(\eta)}) \mathcal{D}A^{(\eta)} \mathcal{D}A_\perp = \Delta_{FP}\, \mathcal{D}A_\perp, \tag{4.19}$$

where $A_\perp$ represents the modes perpendicular to the gauge mode $A^{(\eta)}$:

$$\langle A_\perp, A^{(\eta)} \rangle = 0. \tag{4.20}$$

As such a mode, we can construct one parameter family $A^{(\omega)}$ as follows.

$$\begin{pmatrix} A_2^{(\omega)} \\ A_y^{(\omega)} \end{pmatrix} := \begin{pmatrix} \partial_y d_2 \omega \\ \Delta_0 \omega \end{pmatrix}, \qquad \text{where } \Delta_0 = - *_2 d_2 *_2 d_2. \tag{4.21}$$

This mode gives $\mathcal{S}_{\text{boson}} = \frac{1}{2} \langle A^{(\omega)}, d^\dagger d A^{(\omega)} \rangle$, so from (4.17), we get

$$\int \mathcal{D}A^{(\omega)}\, e^{-\mathcal{S}_{\text{boson}}} = \frac{1}{\Delta_{FP}}. \tag{4.22}$$

Therefore, if we can identify the remaining modes which are perpendicular to both of (4.13) and (4.21), we can forget the effect of Δ_{FP}, and the modes are represented as follows.

$$A_y = 0, \quad *_2 d_2 *_2 A_2 = 0. \tag{4.23}$$

The second condition is equivalent to the Coulomb gauge condition

$$\nabla_i A_2^i = 0, \tag{4.24}$$

where i runs for ϑ, φ. In summary, what we have to consider is the path integral over $(A_i, \sigma \mid \overline{\lambda}, \lambda)$ weighted by the following actions.

$$\mathcal{S}_{\text{boson}}^{gf} = \int dy \int \begin{pmatrix} A_2 \\ \sigma \end{pmatrix}^T \wedge *_2 \begin{pmatrix} - *_2 d_2 *_2 d_2 - \partial_y^2 & - *_2 d_2 \\ *_2 d_2 & - *_2 d_2 *_2 d_2 - \partial_y^2 + 1 \end{pmatrix} \begin{pmatrix} A_2 \\ \sigma \end{pmatrix}, \tag{4.25}$$

$$\mathcal{S}_{\text{fermion}} = \int dy \int \sin\vartheta d\vartheta d\varphi\, \overline{\lambda} \Big(i\gamma^i \nabla_i + i\gamma_3 \Big(\partial_y - \frac{1}{2} \Big) \Big) \lambda, \tag{4.26}$$

constrained by (4.24).

4.1.1 QFT on $S^2 \times S^1_\beta \to$ QM on S^1_β

Now, we take the following eigenfunction expansion:

$$A^i(\vartheta, \varphi, y) = \sum_{j=1}^{\infty} \sum_{m=-j}^{j} V^i_{jm}(\vartheta, \varphi) A_{jm}(y), \tag{4.27}$$

$$\sigma(\vartheta, \varphi, y) = \sum_{j=0}^{\infty} \sum_{m=-j}^{j} Y_{jm}(\vartheta, \varphi) \sigma_{jm}(y), \tag{4.28}$$

$$\lambda(\vartheta, \varphi, y) = \sum_{j=1/2}^{\infty} \sum_{m=-j}^{j} \sum_{\epsilon} \Upsilon^\epsilon_{jm}(\vartheta, \varphi) \lambda^\epsilon_{jm}(y), \tag{4.29}$$

$$\overline{\lambda}(\vartheta, \varphi, y) = \sum_{j=1/2}^{\infty} \sum_{m=-j}^{j} \sum_{\epsilon} \Upsilon^{\epsilon\,\dagger}_{jm}(\vartheta, \varphi) \overline{\lambda}^\epsilon_{jm}(y), \tag{4.30}$$

where $V^i_{jm}, Y_{jm}, \Upsilon^\epsilon_{jm}$ are spherical harmonics with zero monopole $B = 0$ explained in the Appendix B. Then, the actions (4.25) and (4.26) give many-body quantum mechanics:

$$\begin{aligned} S^{gf}_{\text{boson}} &= \sum_{j=1}^{\infty} \sum_{m=-j}^{j} \int dy \, (A_{jm} \ \sigma_{jm}) \begin{pmatrix} -\partial_y^2 + j(j+1) & \sqrt{j(j+1)} \\ \sqrt{j(j+1)} & -\partial_y^2 + j(j+1) + 1 \end{pmatrix} \begin{pmatrix} A_{jm} \\ \sigma_{jm} \end{pmatrix} \\ &+ \int dt \, \sigma_0(-\partial_t^2 + 1)\sigma_0, \end{aligned} \tag{4.31}$$

$$S_{\text{fermion}} = \sum_{j=1/2}^{\infty} \sum_{m=-j}^{j} \int dy \, \left(\overline{\lambda}^-_{jm} \ \overline{\lambda}^+_{jm}\right) \begin{pmatrix} j+\frac{1}{2} & i(\partial_y - \frac{1}{2}) \\ i(\partial_y - \frac{1}{2}) & -(j+\frac{1}{2}) \end{pmatrix} \begin{pmatrix} \lambda^-_{jm} \\ \lambda^+_{jm} \end{pmatrix} \tag{4.32}$$

The periodicity for each sector can be read from the definition of SCI (3.1) and Table 3.1, then, it becomes as follows.

$$A_{jm}(t+\beta) = e^{-(\beta_1-\beta_2)m} A_{jm}(t), \quad \sigma_{jm}(t+\beta) = e^{-(\beta_1-\beta_2)m} \sigma_{jm}(t), \tag{4.33}$$

$$\overline{\lambda}^\epsilon_{jm}(t+\beta) = e^{(-1-m)\beta_1 + m\beta_2} \overline{\lambda}^\epsilon_{jm}(t), \quad \lambda^\epsilon_{jm}(t+\beta) = e^{(+1-m)\beta_1 + m\beta_2} \lambda^\epsilon_{jm}(t). \tag{4.34}$$

We can calculate the contributions from bosons and fermions explicitly as follows.

Bosonic part

$$\int \mathcal{D}A_2 \mathcal{D}\sigma\, e^{-S^{gf}_{\text{boson}}} = \int \prod_{t\in[0,\beta]} \left(d\sigma_0(t) \prod_{j=1}^{\infty} \prod_{m=-j}^{j} dA_{jm}(t) d\sigma_{jm}(t) \right) e^{-S^{gf}_{\text{boson}}}$$
$$= \prod_{\tilde{j}=1}^{\infty} \prod_{\tilde{m}=-\tilde{j}}^{\tilde{j}-1} \frac{1}{\left(2\sinh\frac{\beta\omega_{\tilde{j},-\tilde{m}}}{2}\right)\left(2\sinh\frac{\beta\omega_{\tilde{j}\tilde{m}}}{2}\right)}, \tag{4.35}$$

where

$$\omega_{jm} = \frac{\beta_1 - \beta_2}{\beta} m + j. \tag{4.36}$$

Note that the $\tilde{m}$ in resulting product runs for $(-\tilde{j}) \sim (\tilde{j}-1)$ not $(-\tilde{j}) \sim (\tilde{j})$. One can derive this results as follows. For simplicity let us denote $(\tilde{m}, \tilde{j}) := \prod_{n\in Z} \left(\left[\frac{2\pi}{\beta} n + i \frac{\beta_1-\beta_2}{\beta} \tilde{m} \right]^2 + \tilde{j}^2 \right)$, then the denominator of (4.35) is a square root of products of the following towers:

$$\begin{array}{c} (0,1) \\ (\mp 1, 1), (0,1) \\ (0,2), (\pm 1, 2) \\ (\mp 2, 2), (\mp 1, 2), (0,2) \\ (0,3), (\pm 1, 3), (\pm, 2, 3) \\ (\mp 3, 3), (\mp 2, 3), (\mp 1, 3)(0,3) \\ \cdots \end{array} \tag{4.37}$$

Easily noticed, $(-\tilde{m}, \tilde{j}) = (\tilde{m}, \tilde{j})$, so we get the result after the zeta-function regularization.

Fermionic part

$$\int \mathcal{D}\bar{\lambda}\mathcal{D}\lambda\, e^{-S_{\text{fermion}}} = \int \prod_{t\in[0,\beta]} \left(\prod_{j=1/2}^{\infty} \prod_{m=-j}^{j} d\overline{\lambda}^{+}_{jm}(t) d\overline{\lambda}^{-}_{jm}(t) d\lambda^{+}_{jm}(t) d\lambda^{-}_{jm}(t) \right) e^{-S_{\text{fermion}}}$$
$$= \prod_{\tilde{j}=1}^{\infty} \prod_{\tilde{m}=-\tilde{j}}^{\tilde{j}-1} \left(2\sinh\frac{\beta\omega_{\tilde{j},-\tilde{m}}}{2}\right)\left(2\sinh\frac{\beta\omega_{\tilde{j}\tilde{m}}}{2}\right), \tag{4.38}$$

where ω_{jm} is same one defined in (4.36). Therefore, (4.35) and (4.38) cancel out each other, and we get trivial 1-loop determinant.

$$\int \mathcal{D}A_2\mathcal{D}\sigma\mathcal{D}\overline{\lambda}\mathcal{D}\lambda\; e^{-\mathcal{S}^{gf}_{\text{boson}}-\mathcal{S}_{\text{fermion}}}$$
$$= 1. \tag{4.39}$$

In a later chapter, we will see non-trivial contribution emerges when we consider the theory not on S^2 but RP^2.

4.2 Matter Multiplet

First of all, the matter field in gauge theory is defined by assigning a certain representation of the gauge group. With $U(1)$ gauge group, the matter representation becomes the $U(1)$ charge $\boldsymbol{q} \in Z$ in our situation as explained later.

Locus The matter Lagrangian (3.65) defines the trivial locus.

$$0 = \phi = \psi = F, \quad 0 = \overline{\phi} = \overline{\psi} = \overline{F}. \tag{4.40}$$

So, there is no need for summation for matter sector. And we get the following actions for the fluctuation fields. We omit˜and integrate out the auxiliary fields for simplicity.

$$\mathcal{S}_{\text{boson}} = \int dt \int \sin\vartheta d\vartheta d\varphi \Big(\mathfrak{D}_\mu \overline{\phi} \mathfrak{D}^\mu \phi + \frac{(\boldsymbol{q}B)^2}{2^2} \overline{\phi}\phi - (2\Delta - 1)\overline{\phi}\mathfrak{D}_t\phi - \Delta(\Delta - 1)\overline{\phi}\phi \Big), \tag{4.41}$$

$$\mathcal{S}_{\text{fermion}} \int dt \int \sin\vartheta d\vartheta d\varphi \Big(- i(\overline{\psi}\gamma^\mu \mathfrak{D}_\mu \psi) - i\frac{\boldsymbol{q}B}{2}(\overline{\psi}\psi) - \frac{i(2\Delta - 1)}{2}(\overline{\psi}\gamma_3\psi) \Big), \tag{4.42}$$

where $\mathfrak{D}_\mu$ is the covariant derivative with respect to the locus gauge field (4.3):

$$\mathfrak{D}_i = \nabla_i - i\boldsymbol{q}A_i^{\text{mon}}, \quad (i = \vartheta, \varphi) \tag{4.43}$$

$$\mathfrak{D}_t = \partial_t - i\boldsymbol{q}\frac{\theta}{\beta}. \tag{4.44}$$

The charge $\boldsymbol{q}$ must be in integers in order to make the gauge transformation acting on the matter fields as a single valued function.

4.2.1 *QFT on* $S^2 \times S^1_\beta \to$ *QM on* S^1_β

As performed in the previous section, we expand the component fields as follows:

$$\psi(\vartheta, \varphi, y) = \sum_{j=\frac{|\boldsymbol{q}B|}{2}}^{\infty} \sum_{m=-j}^{j} Y_{\frac{|\boldsymbol{q}B|}{2}, jm}(\vartheta, \varphi)\psi_{jm}(t), \tag{4.45}$$

$$\psi(\vartheta, \varphi, y) = \sum_{j=\frac{|qB|}{2}+1/2}^{\infty} \sum_{m=-j}^{j} \sum_{\epsilon} \Upsilon^{\epsilon}_{\frac{|qB|}{2}, jm}(\vartheta, \varphi)\psi^{\epsilon}_{jm}(t) + \sum_{m=1/2-\frac{|qB|}{2}}^{\frac{|qB|}{2}-1/2} \Upsilon^{0}_{\frac{|qB|}{2}, m}(\vartheta, \varphi)\psi^{0}_{m}(t), \tag{4.46}$$

$$\overline{\phi}(\vartheta, \varphi, y) = \sum_{j=\frac{|qB|}{2}}^{\infty} \sum_{m=-j}^{j} Y^{*}_{\frac{|qB|}{2}, jm}(\vartheta, \varphi)\overline{\phi}_{jm}(t), \tag{4.47}$$

$$\overline{\psi}(\vartheta, \varphi, y) = \sum_{j=\frac{|qB|}{2}+1/2}^{\infty} \sum_{m=-j}^{j} \sum_{\epsilon} \Upsilon^{\epsilon}_{\frac{|qB|}{2}, jm}{}^{\dagger}(\vartheta, \varphi)\overline{\psi}^{\epsilon}_{jm}(t) + \sum_{m=1/2-\frac{|qB|}{2}}^{\frac{|qB|}{2}-1/2} \Upsilon^{0}_{\frac{|qB|}{2}, m}{}^{\dagger}(\vartheta, \varphi)\overline{\psi}^{0}_{m}(t), \tag{4.48}$$

where $Y_{q,jm}$ and $\Upsilon^{\epsilon}_{q,jm}$ are monopole harmonics explained in the Appendix B. Then, the action (4.41) and (4.42) give many-body quantum mechanics:

$$\mathcal{S}_{\text{boson}} = \sum_{j=\frac{|qB|}{2}}^{\infty} \sum_{m=-j}^{j} \int dt\, \overline{\phi}_{jm}(j + \Delta + \mathfrak{D}_t)(j + 1 - \Delta - \mathfrak{D}_t)\phi_{jm} \tag{4.49}$$

$$\begin{aligned} \mathcal{S}_{\text{boson}} = & \sum_{j=\frac{|qB|}{2}+1/2}^{\infty} \sum_{m=-j}^{j} \int dt \begin{pmatrix} \overline{\psi}^{+}_{jm} & \overline{\psi}^{-}_{jm} \end{pmatrix} \begin{pmatrix} -\frac{\sqrt{(2j+1)^2-(qB)^2}}{2} - i\frac{qB}{2} & -i\mathfrak{D}_t - i\frac{2\Delta-1}{2} \\ -i\mathfrak{D}_t - i\frac{2\Delta-1}{2} & +\frac{\sqrt{(2j+1)^2-(qB)^2}}{2} - i\frac{qB}{2} \end{pmatrix} \begin{pmatrix} \psi^{+}_{jm} \\ \psi^{-}_{jm} \end{pmatrix} \\ & - i\frac{B}{|B|} \sum_{m=1/2-\frac{|qB|}{2}}^{\frac{|qB|}{2}-1/2} \int dt\, \overline{\psi}^{0}_{m}\left(j + \Delta + \mathfrak{D}_t\right)\psi^{0}_{m} \end{aligned} \tag{4.50}$$

The periodicity for each factor can be read from the definition of SCI (3.1) and Table 3.1:

$$\phi_{jm}(t + \beta) = e^{(-\Delta-m)\beta_1+m\beta_2+i\mu}\phi_{jm}(t) \tag{4.51}$$

$$\psi_{jm}(t + \beta) = e^{(-\Delta+1-m)\beta_1+m\beta_2+i\mu}\psi_{jm}(t) \tag{4.52}$$

Then contributions from bosons and fermions become as follows.

Bosonic part

$$\int \mathcal{D}\overline{\phi}\mathcal{D}\phi\, e^{-\mathcal{S}_{\text{boson}}} = \prod_{j\geq\frac{|qB|}{2}} \prod_{m=-j}^{j} \frac{1}{\left(2\sinh\frac{\beta\omega^{1}_{jm}}{2}\right)\left(2\sinh\frac{\beta\omega^{2}_{jm}}{2}\right)}, \tag{4.53}$$

where

$$\beta\omega^{1}_{jm} = -i\boldsymbol{q}\theta + (j - m)\beta_1 + (j + \Delta + m)\beta_2 + i\mu, \tag{4.54}$$

$$\beta\omega^{2}_{jm} = -i\boldsymbol{q}\theta - (j + 1 + m)\beta_1 - (j + 1 - \Delta - m)\beta_2 + i\mu. \tag{4.55}$$

Fermionic part

$$\int \mathcal{D}\overline{\psi}\mathcal{D}\psi\, e^{-S_{\text{fermion}}} = \prod_{\tilde{j}\geq\frac{|qB|}{2}} \Big(\prod_{\tilde{m}=-\tilde{j}}^{\tilde{j}-1} 2\sinh\frac{\beta\omega^1_{\tilde{j}\tilde{m}}}{2}\Big)\Big(\prod_{\tilde{m}=-\tilde{j}-1}^{\tilde{j}} 2\sinh\frac{\beta\omega^2_{\tilde{j}\tilde{m}}}{2}\Big), \tag{4.56}$$

First term in fermion contribution looks similar to the first factors of bosonic contribution in (4.53), but lacking the contribution of $m = j$. So this fermionic contribution cancels almost of the bosonic contributions in product containing ω^1_{jm}. Second term in fermionic part looks similar to the second factors in (4.53), there is contributions of $\tilde{m} = -\tilde{j} - 1$ in surplus. So this bosonic contribution cancels almost of the fermionic contributions containing $\omega^2_{\tilde{j}m}$. Therefore, we get the following total contribution.

$$\int \mathcal{D}\overline{\phi}\mathcal{D}\phi\mathcal{D}\overline{\psi}\mathcal{D}\psi\, e^{-S_{\text{boson}}-S_{\text{fermion}}} = \prod_{j\geq\frac{|qB|}{2}} \frac{2\sinh\frac{\beta\omega^2_{j,-j-1}}{2}}{2\sinh\frac{\beta\omega^1_{j,j}}{2}}. \tag{4.57}$$

Another representation In later chapter, we will use more useful representation of (4.57). We can shift the product with respect to j by defining

$$J = j - \frac{|qB|}{2}, \tag{4.58}$$

then

$$(4.57) = \prod_{J=0}^{\infty} \frac{2\sinh\frac{\beta\omega_f^{(J)}}{2}}{2\sinh\frac{\beta\omega_b^{(J)}}{2}}, \tag{4.59}$$

where we define $\beta\omega_f^{(J)}$ and $\beta\omega_b^{(J)}$ as follows

$$\beta\omega_f^{(J)} = i(q\theta - \mu) + 2\beta_2\Big(J + 1 + \frac{|qB|}{2} - \frac{\Delta}{2}\Big), \tag{4.60}$$

$$\beta\omega_b^{(J)} = -i(q\theta - \mu) + 2\beta_2\Big(J + \frac{|qB|}{2} + \frac{\Delta}{2}\Big). \tag{4.61}$$

Here we ignore the overall sign. Now, after simple deformations, we get the following representation.

$$\Big(x^{(1-\Delta)}e^{-iq\theta}\alpha^{-1}\Big)^{\frac{|qB|}{2}} \frac{(e^{-iq\theta}\alpha^{-1}x^{2-\Delta+|qB|}; x^2)_\infty}{(e^{iq\theta}\alpha^{+1}x^{\Delta+|qB|}; x^2)_\infty}, \tag{4.62}$$

where $(z, q)_\infty$ is called *quantum Pochhammer symbol* or *q-shifted factorial* [11] defined by

$$(A; q)_\infty = \prod_{J=0}^{\infty}(1 - Aq^J). \tag{4.63}$$

We used zeta function regularization to get the prefactor here. As one can noticed by comparing it to the calculation of free harmonic oscillator in Chap. 2, this part corresponds to the Casimir energy of the many-body system.

4.3 Formulas

We summarize here formulas to get SCI of our SUSY theories on $S^2 \times S^1_\beta$.

4.3.1 Non Gauge Theory

In this case, we assume that there are $2N_f$ dynamical fields,

$$\Phi_a = (\phi_a, F_a|\psi_a), \quad \overline{\Phi}_a = (\overline{\phi}_a, \overline{F}_a|\overline{\psi}_a), \quad a = 1, \dots, N_f. \tag{4.64}$$

We assign dimension Δ_a and flavor charge $\boldsymbol{f}_a$ to each multiplet, and consider the following action:

$$\mathcal{S}[\Phi, \overline{\Phi}] = \sum_{a=1}^{N_f} \mathcal{S}^{q=0}_{mat}[\Phi_a, \overline{\Phi}_a] + W[\Phi] + \overline{W}[\overline{\Phi}], \tag{4.65}$$

where $\mathcal{S}^{q=0}_{mat}$ is the action (3.65) with $\boldsymbol{q} = 0$. We can take arbitrary superpotential W. The only restriction is that the flavor charge assignments have to preserve W. In this case, the SCI can be obtained just by turning off the B and θ in (4.62) and taking product over N_f contributions:

$$\mathcal{I}(x, \alpha) = \prod_{a=1}^{N_f} \frac{(\alpha^{-f_a} x^{2-\Delta_a}; x^2)_\infty}{(\alpha^{+f_a} x^{\Delta_a}; x^2)_\infty}. \tag{4.66}$$

4.3.2 Gauge Theory

For simplicity, we consider single gauge field (vector multiplet):

$$V = (A_\mu, \sigma, D|\overline{\lambda}, \lambda). \tag{4.67}$$

Of course, we can add charged matter multiplets:

$$\Phi_a = (\phi_a, F_a|\psi_a), \quad \overline{\Phi}_a = (\overline{\phi}_a, \overline{F}_a|\overline{\psi}_a), \quad a = 1, \ldots, N_f, \tag{4.68}$$

with Δ_a,$\boldsymbol{f}_a$ and $U(1)$ charges $\boldsymbol{q}_a$. We assume action as follows.

$$\mathcal{S}[V; \Phi, \overline{\Phi}] = \mathcal{S}_{YM}[V] + \sum_{a=1}^{N_f} \mathcal{S}_{mat}^{\boldsymbol{q}_a}[V; \Phi_a, \overline{\Phi}_a] + W[\Phi] + \overline{W}[\overline{\Phi}], \tag{4.69}$$

where $\mathcal{S}_{YM}$ is the action (3.64) with $U(1)$ gauge group. See [1, 3] for more detail. In this case, we should sum up $B \in Z$ and integrate $\theta \in [0, 2\pi]$ weighted by N_f product of (4.62):

$$\mathcal{I}(x, \alpha) = \sum_{B \in Z} \int_0^{2\pi} \frac{d\theta}{2\pi} \prod_{a=1}^{N_f} \left(x^{(1-\Delta_a)} e^{-i\boldsymbol{q}_a \theta} \alpha^{-\boldsymbol{f}_a} \right)^{\frac{|\boldsymbol{q}_a B|}{2}} \frac{(e^{-i\boldsymbol{q}_a\theta} \alpha^{-\boldsymbol{f}_a} x^{2-\Delta_a+|\boldsymbol{q}_a B|}; x^2)_\infty}{(e^{i\boldsymbol{q}_a\theta} \alpha^{+\boldsymbol{f}_a} x^{\Delta_a+|\boldsymbol{q}_a B|}; x^2)_\infty}. \tag{4.70}$$

References

1. S. Kim, The complete superconformal index for N = 6 Chern-Simons theory. Nucl. Phys. **B821**, 241–284 (2009). arXiv:0903.4172 [hep-th]
2. D. Gang, Chern-Simons theory on L(p, q) lens spaces and Localization. arXiv:0912.4664 [hep-th]
3. Y. Imamura, S. Yokoyama, Index for three dimensional superconformal field theories with general R-charge assignments. J. High Energy Phys. **1104**, 007 (2011). arXiv:1101.0557 [hep-th]
4. A. Kapustin, B. Willett, Generalized superconformal index for three dimensional field theories. arXiv:1106.2484 [hep-th]
5. T. Eguchi, P.B. Gilkey, A.J. Hanson, Gravitation, gauge theories and differential geometry. Phys. Rept. **66**, 213 (1980)
6. C. Nash, S. Sen, Topology and Geometry for Physicists (1983)
7. M. Nakahara, Geometry, topology and physics (2003)
8. N. Hama, K. Hosomichi, S. Lee, SUSY gauge theories on squashed three-spheres. J. High Energy Phys. **1105**, 014 (2011). arXiv:1102.4716 [hep-th]
9. J. Gomis, S. Lee, Exact Kahler potential from gauge theory and mirror symmetry. J. High Energy Phys. **1304**, 019 (2013). arXiv:1210.6022 [hep-th]
10. H. Kim, S. Lee, P. Yi, Exact partition functions on RP^2 and orientifolds. J. High Energy Phys. **1402**, 103 (2014). arXiv:1310.4505 [hep-th]
11. G. Gasper, M. Rahman, *Basic Hypergeometric Series*, vol. 96, 2nd edn. (Cambridge university press, Cambridge, 2004)

Chapter 5
Localization Calculous of SCI with $M^2 = RP^2$

In this chapter, we explain our main results on new SCI by taking $M^2 = RP^2$. The curved space $RP^2 \times S^1_\beta$ can be constructed from $S^2 \times S^1_\beta$ by taking the identification

$$(\pi - \vartheta, \pi + \varphi, y) \sim (\vartheta, \varphi, y). \tag{5.1}$$

As same, a field on $RP^2 \times S^1_\beta$ is defined by imposing boundary condition for the field on $S^2 \times S^1_\beta$ under (5.1), we will call it *parity condition*. However, we cannot take arbitrary parity condition because most of them break the supersymmetry and it spoils the validity for using supersymmetric localization techniques. Therefore, we start our argument from the discussion of the possible *supersymmetric* parity condition which preserves supersymmetry. This simple operation causes non-trivial effects, for example, the localization locus for vector multiplet drastically changes, and the resulting SCIs differ from the ones in Chap. 4.

5.1 Supersymmetric Parity Conditions

As studied in [1] in the context of two-dimensional supersymmetric field theory, we can find supersymmetric parity conditions as follows. Our guiding principles are

- $(\text{parity})^2 = (-1)^F$ where F is the fermion number operator,
- SUSY exact Lagrangians, (3.64) and (3.65), must be invariant under the parity,
- Supersymmetries, δ_ϵ and $\delta_{\bar{\epsilon}}$, must be consistent with the parity.

Let us comment on the second assumption. This requirement is too strong because one should assume parity invariance of not (3.64) or (3.65) alone, but *full* Lagrangian, e.g. (4.69). We will comment on this generic case in Chap. 7.

A. Tanaka, *Superconformal Index on* $\mathrm{RP}^2 \times \mathrm{S}^1$ *and 3D Mirror Symmetry*, Springer Theses, DOI 10.1007/978-981-10-1398-0_5

Vector multiplet We find a set of parity conditions for the vector multiplet as follows.[1]

$$\begin{aligned}
A_\vartheta(\pi-\vartheta,\pi+\varphi,y) &= -A_\vartheta(\vartheta,\varphi,y), \quad A_{\varphi,y}(\pi-\vartheta,\pi+\varphi,y) = +A_{\varphi,y}(\vartheta,\varphi,y),\\
\sigma(\pi-\vartheta,\pi+\varphi,y) &= -\sigma(\vartheta,\varphi,y),\\
\lambda(\pi-\vartheta,\pi+\varphi,y) &= +i\gamma_1\lambda(\vartheta,\varphi,y), \quad \overline{\lambda}(\pi-\vartheta,\pi+\varphi,y) = -i\gamma_1\overline{\lambda}(\vartheta,\varphi,y),\\
D(\pi-\vartheta,\pi+\varphi,y) &= +D(\vartheta,\varphi,y).
\end{aligned} \tag{5.2}$$

One flavor matter multiplet The one flavor matter multiplet has two choices:

$$\begin{aligned}
\phi(\pi-\vartheta,\pi+\varphi,y) &= \pm\phi(\vartheta,\varphi,y), \quad \overline{\phi}(\pi-\vartheta,\pi+\varphi,y) = \pm\overline{\phi}(\vartheta,\varphi,y),\\
\psi(\pi-\vartheta,\pi+\varphi,y) &= \mp i\gamma_1\psi(\vartheta,\varphi,y), \quad \overline{\psi}(\pi-\vartheta,\pi+\varphi,y) = \pm i\gamma_1\overline{\psi}(\vartheta,\varphi,y),\\
F(\pi-\vartheta,\pi+\varphi,y) &= \pm F(\vartheta,\varphi,y), \quad \overline{F}(\pi-\vartheta,\pi+\varphi,y) = \pm\overline{F}(\vartheta,\varphi,y).
\end{aligned} \tag{5.3}$$

Many flavors matter multiplets We use $a, b, \ldots$ as flavor indices $a = 1, \ldots, N_f$, then

$$\begin{aligned}
\phi_a(\pi-\vartheta,\pi+\varphi,y) &= \sum_{b=1}^{N_f} M_{ab}\phi_b(\vartheta,\varphi,y), \quad \overline{\phi}_a(\pi-\vartheta,\pi+\varphi,y) = \sum_{b=1}^{N_f} N_{ab}\overline{\phi}_b(\vartheta,\varphi,y),\\
\psi_a(\pi-\vartheta,\pi+\varphi,y) &= -i\gamma_1\sum_{b=1}^{N_f} M_{ab}\psi_b(\vartheta,\varphi,y), \quad \overline{\psi}_a(\pi-\vartheta,\pi+\varphi,y) = i\gamma_1\sum_{b=1}^{N_f} N_{ab}\overline{\psi}_b(\vartheta,\varphi,y),\\
F_a(\pi-\vartheta,\pi+\varphi,y) &= \sum_{b=1}^{N_f} M_{ab}F_b(\vartheta,\varphi,y), \quad \overline{F}_a(\pi-\vartheta,\pi+\varphi,y) = \sum_{b=1}^{N_f} N_{ab}\overline{F}_b(\vartheta,\varphi,y),
\end{aligned} \tag{5.4}$$

where $(M_{ab})_{a,b=1,\ldots,N_f} = \mathbf{M}$ and $(N_{ab})_{a,b=1,\ldots,N_f} = \mathbf{N}$ are $N_f \times N_f$ matrices constrained by

$$\mathbf{N}^{\mathrm{T}}\mathbf{M} = \mathbf{1}, \quad \mathbf{M}^2 = \mathbf{N}^2 = \mathbf{1}. \tag{5.5}$$

Comments on the parity condition Suppose we have a doublet and the parity condition described by the 2×2 matrices

$$\mathbf{M} = \mathbf{N} = \begin{pmatrix} 0 & 1 \\ 1 & 0 \end{pmatrix}, \tag{5.6}$$

then we can lift its Lagrangian on $RP^2 \times S^1_\beta$ to the one on $S^2 \times S^1_\beta$ by defining a new matter multiplet on $S^2 \times S^1_\beta$ as

[1] There is another possible condition, but we will not explain it here. See [2] for more detail.

$$\Phi(\vartheta,\varphi,y)=\begin{cases}\Phi_1(\vartheta,\varphi,y), & \vartheta\in\left[0,\frac{\pi}{2}\right]\\ \Phi_2(\vartheta,\varphi,y), & \vartheta\in\left[\frac{\pi}{2},\pi\right]\end{cases}. \tag{5.7}$$

The authors of [1] also commented on this fact. This is quite similar to the *doubling trick* in string theory. In Chap. 6, we use such parity condition exactly in the context of three-dimensional mirror symmetry.

5.2 Vector Multiplet

We focus on the gauge theory with abelian gauge field for simplicity as same as in the previous chapter.

Locus Now, let us remind that the Lagrangian (3.64) again,

$$\begin{aligned}\mathcal{L}_{\rm YM}&=\mathcal{F}_\mu\mathcal{F}^\mu+D^2+i\overline{\lambda}\gamma^\mu\mathcal{D}_\mu\lambda-\frac{i}{2}\overline{\lambda}\gamma_3\lambda,\\ \mathcal{F}^\mu&=\frac{1}{2}\epsilon^{\mu\rho\sigma}F_{\rho\sigma}+\partial^\mu\sigma+\delta_3^\mu\sigma.\end{aligned} \tag{5.8}$$

The bosonic terms are obviously positive definite. Therefore, the localization locus is determined by the following equations:

$$0=\mathcal{F}^\mu=D. \tag{5.9}$$

Note that we cannot take the Dirac monopole configuration $A_{\rm mon}$ in (4.3) because it breaks the parity condition (5.2). Instead of it, we can take the flat connection $A^{(\pm)}_{\rm flat}$ on RP^2.

$$A=A^{(\pm)}_{\rm flat}+\frac{\theta}{\beta}dt,\quad \sigma=0, \tag{5.10}$$

where $A^{(\pm)}_{\rm flat}$ represents holonomies of RP^2 along the non-contractible cycle $[\gamma]\neq 0\in\pi_1(RP^2)$. It is also characterized by

$$e^{i\oint_\gamma A^{(\pm)}_{\rm flat}}=\pm 1. \tag{5.11}$$

The constraint on the parameter θ is invariant.

$$\theta\in[0,2\pi]. \tag{5.12}$$

As explained in the end of Chap. 3, in the context of the supersymmetric localization, we expand field V around the locus V_0 which is parametrized by $\pm 1, \theta$:

$$V = V_0[\pm 1, \theta] + \frac{1}{\sqrt{t}}\tilde{V}, \quad \text{where} \quad V_0[\pm 1, \theta] = \left(A^{(\pm)}_{\text{flat}} + \frac{\theta}{\beta}dt, 0, 0 \middle| 0, 0\right) \tag{5.13}$$

and $\tilde{V}$ represents fluctuation. It means that the original path integral is composed from the summation over ± 1 and integration over $\theta \in [0, 2\pi]$, and path integral over the fluctuation $\tilde{V}$:

$$\int \mathcal{D}V e^{-S_{YM}[V]} = \sum_{\pm 1} \int_0^{2\pi} \frac{d\theta}{2\pi} \int \mathcal{D}\tilde{V} e^{-\tilde{S}_{YM}[\tilde{V}]}. \tag{5.14}$$

Note that there is no monopole but ± 1 holonomies, so the summation is not infinite summation over the integers but constructed of just 2 terms, $+1$ sector and -1 sector.

5.2.1 QFT on $RP^2 \times S^1_\beta \to$ QM on S^1_β

The gauge fixing procedure in the previous chapter also works on $RP^2 \times S^1_\beta$, so we can use the Lagrangians

$$S^{gf}_{\text{boson}} = \int dy \int \begin{pmatrix} A_2 \\ \sigma \end{pmatrix}^T \wedge *_2 \begin{pmatrix} - *_2 d_2 *_2 d_2 - \partial_y^2 & - *_2 d_2 \\ *_2 d_2 & - *_2 d_2 *_2 d_2 - \partial_y^2 + 1 \end{pmatrix} \begin{pmatrix} A_2 \\ \sigma \end{pmatrix}, \tag{5.15}$$

$$S_{\text{fermion}} = \int dy \int \sin\vartheta d\vartheta d\varphi \, \overline{\lambda}\left(i\gamma^i \nabla_i + i\gamma_3 \left(\partial_y - \frac{1}{2}\right)\right)\lambda, \tag{5.16}$$

constrained by (4.24). One might think that the naive expansion of each field with respect to the harmonics V^i_{jm}, Ψ^ϵ_{jm}, Y_{jm} works. However it is not. Precisely speaking, the range of summation for j is constrained because of the parity condition (5.2). As one can find in the Appendix of [1], each harmonics behaves as follows[2]:

$$Y_{jm}(\pi - \vartheta, \pi + \varphi) = (-1)^j Y_{jm}(\vartheta, \varphi), \tag{5.17}$$

$$\Psi^\pm_{jm}(\pi - \vartheta, \pi + \varphi) = \mp i(-1)^{j-\frac{1}{2}}\gamma_1 \Psi^\pm_{jm}(\vartheta, \varphi), \tag{5.18}$$

$$V_{jm}(\pi - \vartheta, \pi + \varphi) = (-1)^{j+1} V_{jm}(\vartheta, \varphi). \tag{5.19}$$

[2] Our V_{jm} corresponds to C^2_{jm} in their notation.

We have no fermion zero mode, and we take eigenspinor Ψ for a modified Dirac operator $-i\gamma_3\gamma^i\mathfrak{D}_i$ rather than Υ for the Dirac operator $-i\gamma^i\mathfrak{D}_i$. V_{jm} is the 1-form constructed by $(V_{jm})_\vartheta d\vartheta + (V_{jm})_\varphi d\varphi$. The harmonics which preserves supersymmetric parity conditions in (5.2) only contribute to the expansion as follows.

$$A^i(\vartheta,\varphi,y) = \sum_{\substack{j=2k+1\\k\geq 0}} \sum_{m=-j}^{j} V^i_{jm}(\vartheta,\varphi)A_{jm}(y), \tag{5.20}$$

$$\sigma(\vartheta,\varphi,y) = \sum_{\substack{j=2k+1\\k\geq 0}} \sum_{m=-j}^{j} Y_{jm}(\vartheta,\varphi)\sigma_{jm}(y), \tag{5.21}$$

$$\lambda(\vartheta,\varphi,y) = \sum_{\substack{j=2k+1/2\\k\geq 0}} \sum_{m=-j}^{j} \Psi^-_{jm}(\vartheta,\varphi)\lambda^-_{jm}(y) + \sum_{\substack{j=2k+3/2\\k\geq 0}} \sum_{m=-j}^{j} \Psi^+_{jm}(\vartheta,\varphi)\lambda^+_{jm}(y), \tag{5.22}$$

$$\overline{\lambda}(\vartheta,\varphi,y) = \sum_{\substack{j=2k+1/2\\k\geq 0}} \sum_{m=-j}^{j} \overline{\Psi}^-_{jm}(\vartheta,\varphi)\overline{\lambda}^-_{jm}(y) + \sum_{\substack{j=2k+3/2\\k\geq 0}} \sum_{m=-j}^{j} \overline{\Psi}^+_{jm}(\vartheta,\varphi)\overline{\lambda}^+_{jm}(y). \tag{5.23}$$

Then, the actions (5.15) and (5.16) give many-body quantum mechanics defined by the following actions:

$$S^{gf}_{\text{boson}} = \sum_{\substack{j=2k+1\\k\geq 0}} \sum_{m=-j}^{j} \int dy\ \begin{pmatrix} A_{jm} & \sigma_{jm} \end{pmatrix} \begin{pmatrix} -\partial_y^2 + j(j+1) & \sqrt{j(j+1)} \\ \sqrt{j(j+1)} & -\partial_y^2 + j(j+1) + 1 \end{pmatrix} \begin{pmatrix} A_{jm} \\ \sigma_{jm} \end{pmatrix}, \tag{5.24}$$

$$\begin{aligned} S_{\text{fermion}} = i & \sum_{\substack{j=2k+1/2\\k\geq 0}} \sum_{m=-j}^{j} \int dy\, \overline{\lambda}^-_{jm} \left(\left(j + \frac{1}{2} \right) + \left(\partial_t - \frac{1}{2} \right) \right) \lambda^-_{jm} \\ + i & \sum_{\substack{j=2k+3/2\\k\geq 0}} \sum_{m=-j}^{j} \int dy\, \overline{\lambda}^+_{jm} \left(-\left(j + \frac{1}{2} \right) + \left(\partial_t - \frac{1}{2} \right) \right) \lambda^+_{jm}. \end{aligned} \tag{5.25}$$

The periodicity for each field can be read from the definition of SCI (3.1) and Table 3.1, then it becomes as

$$A_{jm}(t+\beta) = e^{-(\beta_1-\beta_2)m} A_{jm}(t), \quad \sigma_{jm}(t+\beta) = e^{-(\beta_1-\beta_2)m}\sigma_{jm}(t) \tag{5.26}$$

$$\overline{\lambda}^\epsilon_{jm}(t+\beta) = e^{(-1-m)\beta_1+m\beta_2}\overline{\lambda}^\epsilon_{jm}(t), \quad \lambda^\epsilon_{jm}(t+\beta) = e^{(+1-m)\beta_1+m\beta_2}\lambda^\epsilon_{jm}(t). \tag{5.27}$$

Therefore, we get each contribution as follows.

Bosonic part

$$\begin{aligned}\int \mathcal{D}A_2\mathcal{D}\sigma\, e^{-S^{gf}_{\text{boson}}} &= \int \prod_{t\in[0,\beta]} \Big(\prod_{\substack{j=2k+1\\k\geq 0}} \prod_{m=-j}^{j} dA_{jm}(t)d\sigma_{jm}(t)\Big)\, e^{-S^{gf}_{\text{boson}}} \\ &= \prod_{\substack{j=2k+1\\k\geq 0}} \prod_{m=-j}^{j} \frac{1}{\left(2\sinh\frac{\omega_{jm}}{2}\right)\left(2\sinh\frac{\omega_{j+1,m}}{2}\right)},\end{aligned} \tag{5.28}$$

where

$$\omega_{jm} = \frac{\beta_1 - \beta_2}{\beta} m + j. \tag{5.29}$$

Fermionic part

$$\begin{aligned}&\int \mathcal{D}\overline{\lambda}\mathcal{D}\lambda\, e^{-S_{\text{fermion}}} \\ &= \prod_{t\in[0,\beta]} \left(\prod_{\substack{j=2k+1/2\\k\geq 0}}^{\infty} \prod_{m=-j}^{j} d\lambda^{-}_{jm}(t)d\overline{\lambda^{-}_{jm}}(t) \right)\left(\prod_{\substack{j=2k+3/2\\k\geq 0}}^{\infty} \prod_{m=-j}^{j} d\lambda^{+}_{jm}(t)d\overline{\lambda^{+}_{jm}}(t) \right) e^{-S_{\text{fermion}}} \\ &= \prod_{\substack{j=2k+1\\k\geq 0}} \left(\prod_{m=-j+1}^{j} 2\sinh\frac{\beta\omega_{jm}}{2} \right)\left(\prod_{m=-j-1}^{j} 2\sinh\frac{\beta\omega_{j+1,m}}{2} \right).\end{aligned} \tag{5.30}$$

Note that, in contrast to the case of $M^2 = S^2$ (4.39), we get the following non-trivial contribution even from the vector multiplet.

$$\begin{aligned}&\int \mathcal{D}A_2\mathcal{D}\sigma\mathcal{D}\overline{\lambda}\mathcal{D}\lambda\, e^{-S^{gf}_{\text{boson}} - S_{\text{fermion}}} \\ &= \prod_{\substack{j=2k+1\\k\geq 0}} \frac{2\sinh\frac{\beta\omega_{j+1,-(j+1)}}{2}}{2\sinh\frac{\beta\omega_{j,-j}}{2}} = x^{\frac{1}{4}}\frac{(x^4;x^4)_\infty}{(x^2;x^4)_\infty}.\end{aligned} \tag{5.31}$$

The Z_2 action to boson is different from that of fermion because of the difference of their spins. Therefore, the survived modes are different, and as a result, the complete cancellation in (4.39) breaks up.

5.3 Matter Multiplet

Locus The matter Lagrangian (3.65) defines the trivial field contents:

$$0 = \phi = \psi = F, \quad 0 = \overline{\phi} = \overline{\psi} = \overline{F}. \tag{5.32}$$

$$\mathcal{S}_{\text{boson}} = \int dt \int \sin\vartheta d\vartheta d\varphi \Big(D_\mu \overline{\phi} D^\mu \phi - (2\Delta - 1)\overline{\phi} D_t \phi - \Delta(\Delta - 1)\overline{\phi}\phi \Big), \tag{5.33}$$

$$\mathcal{S}_{\text{fermion}} = \int dt \int \sin\vartheta d\vartheta d\varphi \Big(- i(\overline{\psi}\gamma^\mu D_\mu \psi) - \frac{i(2\Delta - 1)}{2}(\overline{\psi}\gamma_3\psi) \Big), \tag{5.34}$$

where D_μ represent the covariant derivative with respect to the locus gauge field (5.10):

$$D_i = \nabla_i - i\boldsymbol{q} A_i^{\text{flat}} \quad (i = \vartheta, \varphi), \tag{5.35}$$

$$D_t = \mathfrak{D}_t = \partial_t - i\boldsymbol{q}\frac{\theta}{\beta}. \tag{5.36}$$

5.3.1 *QFT on* $RP^2 \times S^1_\beta \to$ *QM on* S^1_β

Here, for simplicity, we focus on the following two cases.

One-flavor matter multiplet First, we treat the $e^{i \oint_\gamma \boldsymbol{q} A_{\text{flat}}} = +1$ case in (5.3). In this case, we have to restrict j as follows:

$$\phi(\vartheta, \varphi, y) = \sum_{\substack{j=2k \\ k\geq 0}}^{\infty} \sum_{m=-j}^{j} e^{i\int^x \boldsymbol{q} A_{\text{flat}}} Y_{jm}(\vartheta, \varphi)\phi_{jm}(t), \tag{5.37}$$

$$\psi(\vartheta, \varphi, y) = \sum_{\substack{j=2k+1/2 \\ k>0}} \sum_{m=-j}^{j} e^{i\int^x \boldsymbol{q} A_{\text{flat}}} \Psi^+_{jm}(\vartheta, \varphi)\psi^+_{jm}(y) + \sum_{\substack{j=2k+3/2 \\ k>0}} \sum_{m=-j}^{j} e^{i\int^x \boldsymbol{q} A_{\text{flat}}} \Psi^-_{jm}(\vartheta, \varphi)\psi^-_{jm}(y), \tag{5.38}$$

$$\overline{\phi}(\vartheta, \varphi, y) = \sum_{\substack{j=2k \\ k\geq 0}}^{\infty} \sum_{m=-j}^{j} e^{-i\int^x \boldsymbol{q} A_{\text{flat}}} Y^*_{jm}(\vartheta, \varphi)\overline{\phi}_{jm}(t), \tag{5.39}$$

$$\overline{\psi}(\vartheta, \varphi, y) = \sum_{\substack{j=2k+1/2 \\ k\geq 0}} \sum_{m=-j}^{j} e^{-i\int^x \boldsymbol{q} A_{\text{flat}}} \overline{\Psi}^+_{jm}(\vartheta, \varphi)\overline{\psi}^+_{jm}(y) + \sum_{\substack{j=2k+3/2 \\ k\geq 0}} \sum_{m=-j}^{j} e^{-i\int^x \boldsymbol{q} A_{\text{flat}}} \overline{\Psi}^-_{jm}(\vartheta, \varphi)\overline{\psi}^-_{jm}(y), \tag{5.40}$$

where Y_{jm} and $\Psi^{\pm}_{jm}$ are harmonices explained in the Appendix B. Then, each action (5.33) and (5.34) gives many-body quantum mechanics:

$$\mathcal{S}_{\text{boson}} = \sum_{\substack{j=2k \\ k\geq 0}} \sum_{m=-j}^{j} \int dt\, \overline{\phi}_{jm}(j+\Delta+\mathfrak{D}_t)(j+1-\Delta-\mathfrak{D}_t)\phi_{jm}, \tag{5.41}$$

$$\begin{aligned} \mathcal{S}_{\text{boson}} = i & \sum_{\substack{j=2k+1/2 \\ k\geq 0}} \sum_{m=-j}^{j} \int dt\, \overline{\psi}^{+}_{jm} \left(\left(j+\frac{1}{2} \right) - \left(\mathfrak{D}_t + \frac{2\Delta-1}{2} \right) \right) \psi^{+}_{jm} \\ & + i \sum_{\substack{j=2k+3/2 \\ k\geq 0}} \sum_{m=-j}^{j} \int dt\, \overline{\psi}^{-}_{jm} \left(- \left(j+\frac{1}{2} \right) - \left(\mathfrak{D}_t + \frac{2\Delta-1}{2} \right) \right) \psi^{-}_{jm}. \end{aligned} \tag{5.42}$$

The periodicities can be read from the definition of SCI (3.1) and Table 3.1:

$$\phi_{jm}(t+\beta) = e^{(-\Delta-m)\beta_1+m\beta_2+i\mu}\phi_{jm}(t) \tag{5.43}$$

$$\psi_{jm}(t+\beta) = e^{(-\Delta+1-m)\beta_1+m\beta_2+i\mu}\psi_{jm}(t) \tag{5.44}$$

Then each contribution becomes as follows.

Bosonic part

$$\begin{aligned} \int \mathcal{D}\phi\mathcal{D}\overline{\phi}\, e^{-\mathcal{S}_{\text{boson}}} &= \int \prod_{t\in[0,\beta]} \Big(\prod_{\substack{j=2k \\ k\geq 0}} \prod_{m=-j}^{j} d\overline{\phi}_{jm}(t) d\phi_{jm}(t) \Big)\, e^{-\mathcal{S}_{\text{boson}}} \\ &= \prod_{\substack{j=2k \\ k\geq 0}} \prod_{m=-j}^{j} \frac{1}{\left(2\sinh\frac{\beta\omega^1_{jm}}{2}\right)\left(2\sinh\frac{\beta\omega^2_{jm}}{2}\right)}, \end{aligned} \tag{5.45}$$

where

$$\beta\omega^1_{jm} = -i\boldsymbol{q}\theta + (j-m)\beta_1 + (j+\Delta+m)\beta_2 + i\mu, \tag{5.46}$$

$$\beta\omega^2_{jm} = -i\boldsymbol{q}\theta - (j+1+m)\beta_1 - (j+1-\Delta-m)\beta_2 + i\mu. \tag{5.47}$$

Fermionic part

$$\int \mathcal{D}\overline{\psi}\mathcal{D}\psi\, e^{-S_{\text{fermion}}}$$
$$= \int \prod_{t\in[0,\beta]} \left(\prod_{\substack{j=2k+1/2\\k\geq 0}} \prod_{m=-j}^{j} d\overline{\psi}^{+}_{jm}(t) d\psi^{+}_{jm}(t) \right) \left(\prod_{\substack{j=2k+3/2\\k\geq 0}} \prod_{m=-j}^{j} d\overline{\psi}^{-}_{jm}(t) d\psi^{-}_{jm}(t) \right) e^{-S_{\text{fermion}}}$$
$$= \prod_{\substack{j=2k\\k\geq 0}} \prod_{m=-j-1}^{j} \left(2\sinh\frac{\beta\omega^2_{jm}}{2} \right) \times \prod_{\substack{j=2k+2\\k\geq 0}} \prod_{m=-j}^{j-1} \left(2\sinh\frac{\beta\omega^1_{jm}}{2} \right). \tag{5.48}$$

Then, in total, we get

$$\int \mathcal{D}\overline{\phi}\mathcal{D}\phi\mathcal{D}\overline{\psi}\mathcal{D}\psi\, e^{-S_{\text{boson}}-S_{\text{fermion}}} = \prod_{\substack{j=2k\\k\geq 0}} \frac{2\sinh\frac{\beta\omega^1_{j,-j-1}}{2}}{2\sinh\frac{\beta\omega^2_{j,j}}{2}}$$
$$= x^{+\frac{\Delta-1}{4}} e^{+\frac{i}{4}q\theta} \alpha^{+\frac{1}{4}f} \frac{(e^{-iq\theta}\alpha^{-f} x^{(2-\Delta)}; x^4)_\infty}{(e^{+iq\theta}\alpha^{+f} x^{\Delta}; x^4)_\infty}. \tag{5.49}$$

Now, we turn to the contribution for $e^{i\oint_\gamma qA_{\text{flat}}} = -1$ sector. The only difference is the range for j in bosonic sector. After repeating similar procedure, we get the following contribution.

$$\int \mathcal{D}\overline{\phi}\mathcal{D}\phi\mathcal{D}\overline{\psi}\mathcal{D}\psi\, e^{-S_{\text{boson}}-S_{\text{fermion}}} = x^{-\frac{\Delta-1}{4}} e^{-\frac{i}{4}q\theta} \alpha^{-\frac{1}{4}f} \frac{(e^{-iq\theta}\alpha^{-f} x^{(4-\Delta)}; x^4)_\infty}{(e^{+iq\theta}\alpha^{+f} x^{(2+\Delta)}; x^4)_\infty}. \tag{5.50}$$

Two-flavor matter multiplets with (5.6)**-type parity matrix.**
In this case, as we have noted in (5.7), we can construct one-flavor matter multiplet on $S^2 \times S^1_\beta$ with zero monopole, therefore we easily get the result from (4.62).

$$\int_{RP^2\times S^1_\beta} [\mathcal{D}\overline{\phi}_1\mathcal{D}\phi_1\mathcal{D}\overline{\psi}_1\mathcal{D}\psi_1]\,[\mathcal{D}\overline{\phi}_2\mathcal{D}\phi_2\mathcal{D}\overline{\psi}_2\mathcal{D}\psi_2]\, e^{-S_{\text{boson1}}-S_{\text{fermion1}}-S_{\text{boson2}}-S_{\text{fermion2}}}$$
$$= \int_{S^2\times S^1_\beta} \mathcal{D}\overline{\phi}\mathcal{D}\phi\mathcal{D}\overline{\psi}\mathcal{D}\psi\, e^{-S_{\text{boson}}-S_{\text{fermion}}}$$
$$= \frac{(e^{-iq\theta}\alpha^{-f} x^{2-\Delta}; x^2)_\infty}{(e^{iq\theta}\alpha^{+f} x^{\Delta}; x^2)_\infty}. \tag{5.51}$$

5.4 Formulas

We summarize here the formulas for making SCI of our SUSY theories on $RP^2 \times S^1_\beta$ focusing on the theories composed by multiple of two types matter multiplets discussed in previous section.

5.4.1 Non Gauge Theory

Let us turn on the following dynamical fields.

$$\Phi_a = (\phi_a, F_a|\psi_a), \quad \overline{\Phi}_a = (\overline{\phi}_a, \overline{F}_a|\overline{\psi}_a), \quad a = 1, \ldots, N_f^{\text{single}} \quad \text{with } +1 \text{ in (5.1.2)}, \tag{5.52}$$

$$\Phi^A_{1,2} = (\phi^A_{1,2}, F^A_{1,2}|\psi^A_{1,2}), \quad \overline{\Phi}^A_{1,2} = (\overline{\phi}^A_{1,2}, \overline{F}^A_{1,2}|\overline{\psi}^A_{1,2}), \quad A = 1, \ldots, N_f^{\text{double}} \quad \text{with (5.1.5) in (5.1.3)}. \tag{5.53}$$

We assign dimension Δ_a, Δ_A and flavor charge $\boldsymbol{f}_a, \boldsymbol{f}_A$ to each multiplet, and consider the following action:

$$\mathcal{S}[\Phi, \overline{\Phi}] = \sum_{a=1}^{N_f^{\text{single}}} \mathcal{S}^{q=0}_{mat}[\Phi_a, \overline{\Phi}_a] + \sum_{A=1}^{N_f^{\text{double}}} \mathcal{S}^{q=0}_{mat}[\Phi^A_{1,2}, \overline{\Phi}^A_{1,2}] + W[\Phi] + \overline{W}[\overline{\Phi}], \tag{5.54}$$

where $\mathcal{S}^{q=0}_{mat}$ is the action (3.65) with $\boldsymbol{q} = 0$. We can take arbitrary superpotential W if it is invariant under the parity conditions. The flavor charge assignments $\boldsymbol{f}_a, \boldsymbol{f}_A$ have to preserve W. In this case, we get

$$\mathcal{I}(x, \alpha) = \prod_{a=1}^{N_f^{\text{single}}} x^{+\frac{\Delta_a - 1}{4}} \alpha^{+\frac{1}{4}\boldsymbol{f}_a} \frac{(\alpha^{-\boldsymbol{f}_a} x^{(2-\Delta_a)}; x^4)_\infty}{(\alpha^{+\boldsymbol{f}_a} x^{\Delta_a}; x^4)_\infty} \prod_{A=1}^{N_f^{\text{double}}} \frac{(\alpha^{-\boldsymbol{f}_A} x^{(2-\Delta_A)}; x^2)_\infty}{(\alpha^{+\boldsymbol{f}_A} x^{\Delta_A}; x^2)_\infty} \tag{5.55}$$

5.4.2 Gauge Theory

We consider the $U(1)$ gauge theory with single gauge field (vector multiplet):

$$V = (A_\mu, \sigma, D|\overline{\lambda}, \lambda). \tag{5.56}$$

and the following matter singlets:

$$\Phi_a = (\phi_a, F_a|\psi_a), \quad \overline{\Phi}_a = (\overline{\phi}_a, \overline{F}_a|\overline{\psi}_a), \quad a = 1, \dots, N_f \quad \text{with } +1 \text{ in (5.1.2)}, \tag{5.57}$$

with Δ_a, $\boldsymbol{f}_a$ and $U(1)$ charges $\boldsymbol{q}_a$. The action is

$$\mathcal{S}[V;\Phi,\overline{\Phi}] = \mathcal{S}_{YM}[V] + \sum_{a=1}^{N_f} \mathcal{S}_{mat}^{\boldsymbol{q}_a}[V;\Phi_a,\overline{\Phi}_a] + W[\Phi] + \overline{W}[\overline{\Phi}], \tag{5.58}$$

where $\mathcal{S}_{YM}$ is the action (3.64) with $U(1)$ gauge group. See [3] for more detail. We have to sum up all locus contributions. It means that we should sum up $\pm$ sector's contributions and integrate $\theta \in [0, 2\pi]$. The formula is

$$\begin{aligned}\mathcal{I}(x,\alpha) = &\int_0^{2\pi} \frac{d\theta}{2\pi} \prod_{a=1}^{N_f} x^{+\frac{\Delta_a - 1}{4}} e^{+\frac{i}{4}\boldsymbol{q}\theta} \alpha^{+\frac{1}{4}\boldsymbol{f}_a} \frac{(e^{-i\boldsymbol{q}\theta}\alpha^{-\boldsymbol{f}_a} x^{(2-\Delta_a)}; x^4)_\infty}{(e^{+i\boldsymbol{q}\theta}\alpha^{+\boldsymbol{f}_a} x^{\Delta_a}; x^4)_\infty} \times x^{\frac{1}{4}} \frac{(x^4;x^4)_\infty}{(x^2;x^4)_\infty} \\ &+ \int_0^{2\pi} \frac{d\theta}{2\pi} \prod_{a=1}^{N_f} x^{-\frac{\Delta_a - 1}{4}} e^{-\frac{i}{4}\boldsymbol{q}\theta} \alpha^{-\frac{1}{4}\boldsymbol{f}_a} \frac{(e^{-i\boldsymbol{q}\theta}\alpha^{-\boldsymbol{f}_a} x^{(4-\Delta_a)}; x^4)_\infty}{(e^{+i\boldsymbol{q}\theta}\alpha^{+\boldsymbol{f}_a} x^{(2+\Delta_a)}; x^4)_\infty} \times x^{\frac{1}{4}} \frac{(x^4;x^4)_\infty}{(x^2;x^4)_\infty}.\end{aligned} \tag{5.59}$$

References

1. H. Kim, S. Lee, P. Yi, Exact partition functions on RP^2 and orientifolds. J. High Energy Phys. **1402**, 103 (2014). arXiv:1310.4505 [hep-th]
2. A. Tanaka, H. Mori, T. Morita, Abelian 3d mirror symmetry on $RP^2 \times S^1$ with $N_f = 1$. J. High Energy Phys. **09**, 154 (2015). arXiv:1505.07539 [hep-th]
3. A. Tanaka, H. Mori, T. Morita, Superconformal index on $RP^2 \times S^1$ and mirror symmetry. Phys. Rev. D **91**, 105023 (2015). arXiv:1408.3371 [hep-th]

Chapter 6
An Application: Three-Dimensional Abelian Mirror Symmetry

In this chapter, we apply the exact results of SCI to the check of a conjectural duality called three-dimensional mirror symmetry [1–3], duality between Supersymmetric Quantum ElectroDynamics (SQED) and XYZ-model.

6.1 Duality Between SQED and XYZ-Model

First, let us survey each theory's Lagrangian, global symmetries, etc.

6.1.1 XYZ-Model

Degrees of freedom This is a non gauge theory constructed of three matter multiplets

$$X = (\phi_X, F_X, |\psi_X), \quad Y = (\phi_Y, F_Y, |\psi_Y), \quad Z = (\phi_Z, F_Z, |\psi_Z), \quad \text{+their conjugates.} \tag{6.1}$$

Dimensions Each multiplet have the following dimensions controlled by Δ:

$$\Delta_X = \Delta_Y = 1 - \Delta, \quad \Delta_Z = 2\Delta. \tag{6.2}$$

Lagrangian Lagrangian contains superpotential term in the form of XYZ.

$$\begin{aligned} &\mathcal{S}_{XYZ}[X, Y, Z] \\ &= \mathcal{S}^{q=0}_{mat}[X] + \mathcal{S}^{q=0}_{mat}[Y] + \mathcal{S}^{q=0}_{mat}[Z] + \int dx^3 (XYZ)|_{\theta\theta} + \int dx^3 (\overline{XYZ})|_{\overline{\theta\theta}} \end{aligned} \tag{6.3}$$

A. Tanaka, *Superconformal Index on* RP^2 × S^1 *and 3D Mirror Symmetry*,
Springer Theses, DOI 10.1007/978-981-10-1398-0_6

Global symmetries There are two global symmetries called $U(1)_V$ and $U(1)_A$. We denote here the corresponding flavor charges as f_V, f_A.

	X	Y	Z
f_V	$+1$	-1	0
f_A	$+1$	$+1$	-2

(6.4)

Parameters of the vacua As well known, scalars can take vacuum expectation values (VEVs). In this case there are three scalars. Therefore, the parameters of the vacua are the following three VEVs:

$$\langle \phi_X \rangle, \quad \langle \phi_Y \rangle, \quad \langle \phi_Z \rangle. \tag{6.5}$$

6.1.2 SQED

Degrees of freedom This is a gauge theory constructed from one vector multiplet and two charged matter multiplets.

$$V = (A_\mu, \sigma, D|\overline{\lambda}, \lambda), \tag{6.6}$$

$$Q = (\phi_Q, F_Q, |\psi_Q), \quad \tilde{Q} = (\phi_{\tilde{Q}}, F_{\tilde{Q}}, |\psi_{\tilde{Q}}), \quad +\text{their conjugates.} \tag{6.7}$$

Q has a charge $+1$, and $\tilde{Q}$ has a charge -1 under the $U(1)$ gauge symmetry.

Dimensions Each multiplet has the following dimensions:

$$\Delta_Q = \Delta_{\tilde{Q}} = \Delta. \tag{6.8}$$

Dual photon In 3 dimension, degrees of freedom of the massless vector is equivalent to that of a real scalar ρ through the following equation:

$$\frac{1}{2}\epsilon_{\mu\nu\rho} F^{\nu\rho} = \partial_\mu \rho. \tag{6.9}$$

The real scalar field ρ is called *dual photon*.

Lagrangian Lagrangian is as follows.

$$S_{SQED}[V, Q, \tilde{Q}] = S_{YM}[V] + S_{mat}^{q=+1}[V; Q] + S_{mat}^{q=-1}[V; \tilde{Q}]. \tag{6.10}$$

Global symmetries There are two global symmetries called $U(1)_J$ and $U(1)_A$. We denote here the corresponding flavor charges as $\tilde{f}_J, \tilde{f}_A$.

$$
\begin{array}{c|c|c|c|c}
 & e^{\sigma+i\rho} & e^{-(\sigma+i\rho)} & Q & \tilde{Q} \\
\hline
\tilde{f}_J & +1 & -1 & 0 & 0 \\
\hline
\tilde{f}_A & 0 & 0 & +1 & +1
\end{array}
\tag{6.11}
$$

Parameters of the vacua The scalar VEV have to preserve the gauge symmetry, so the meson field, the lowest component of $\tilde{Q}Q$ is one of the good coordinates. The other ones are $e^{\sigma \pm i\rho}$. Therefore, there are three relevant VEVs.

$$
\langle e^{\sigma+i\rho} \rangle, \quad \langle e^{-(\sigma+i\rho)} \rangle, \quad \langle \phi_{\tilde{Q}} \phi_Q \rangle. \tag{6.12}
$$

6.2 Check in $M^2 = S^2$ Case

At the beginning of the discovery of this duality, there were some indirect checks, moduli space equivalence, parity anomaly matching, etc. [1, 2]. After the developments of the exact calculation based on localization techniques, we can see its duality in the form of mathematical identity. For example, through the sphere partition function Z, the equivalence $Z_{XYZ} = Z_{SQED}$ reduces to the identity [4, 5]

$$
\frac{1}{\cosh \frac{p}{2}} = \int_{-\infty}^{\infty} dx \frac{e^{ipx}}{\cosh \pi x}. \tag{6.13}
$$

This is the Fourier transformation of the $\cosh^{-1}$ function. In this chapter, we review recent developments of the precision check of the duality by using superconformal index on $S^2 \times S^1_\beta$. In this chapter, for simplicity, we turn on only the fugacity for $U(1)_A$ global symmetries.

6.2.1 SCI of XYZ-Model

According to the formula in (4.66) and the charge assignments in (6.4), we get

$$
\mathcal{I}^{\Delta}_{\text{XYZ}}(x, \alpha) = \left(\frac{(\alpha^{-1} x^{(1+\Delta)}; x^2)_\infty}{(\alpha^{+1} x^{(1-\Delta)}; x^2)_\infty} \right)^2 \frac{(\alpha^{+2} x^{2(1-\Delta)}; x^2)_\infty}{(\alpha^{-2} x^{2\Delta}; x^2)_\infty}. \tag{6.14}
$$

For example, we can expand it with respect to x by taking spatial values for $\Delta = 1/2, \alpha = 1$ as follows

$$
\mathcal{I}^{1/2}_{\text{XYZ}}(x, 1) = 1 + 2x^{1/2} + 3x + 2x^{3/2} + x^2 + 2x^{5/2} + 4x^3 + 4x^{7/2} - 2x^{9/2} \ldots \tag{6.15}
$$

This means that there are infinitely many BPS states (3.3) summarized as follows.

$\hat{H}+\hat{j}^3$	0	1/2	1	3/2	2	5/2	3	7/2	9/2	...
$\#_b - \#_f$ in BPS states	1	2	3	2	1	2	4	4	−2	...

(6.16)

6.2.2 SCI of SQED

According to the formula (4.70) and the charge assignments in (6.11), we get[1]

$$\mathcal{I}^{\Delta}_{\text{SQED}}(x,\alpha^{-1}) = \sum_{B\in Z}\int_0^{2\pi}\frac{d\theta}{2\pi}\left(x^{(1-\Delta)}\alpha\right)^{|B|}\frac{(e^{-i\theta}\alpha x^{2-\Delta+|B|};x^2)_\infty}{(e^{i\theta}\alpha^{-1}x^{\Delta+|B|};x^2)_\infty}\times\frac{(e^{i\theta}\alpha x^{2-\Delta+|B|};x^2)_\infty}{(e^{-i\theta}\alpha^{-1}x^{\Delta+|B|};x^2)_\infty}. \tag{6.17}$$

By using mathematica, we can get numerical value for $\Delta=1/2$, $\alpha=1$ as follows:

$$\mathcal{I}^{1/2}_{\text{SQED}}(x,1) = 1+2x^{1/2}+3x+2x^{3/2}+x^2+2x^{5/2}+4x^3+4x^{7/2}-2x^{9/2}+\dots \tag{6.18}$$

This looks in agreement with (6.15). In fact, one can find the analytic proof of

$$\mathcal{I}^{\Delta}_{\text{XYZ}}(x,\alpha) = \mathcal{I}^{\Delta}_{\text{SQED}}(x,\alpha^{-1}), \tag{6.19}$$

in Appendix C.1.

6.2.3 Check in $M^2 = RP^2$ Case

We can also check the duality through SCI on $RP^2\times S^1_\beta$ [6]. This case, we have to identify supersymmetric parity conditions in each side. The hint for it is the correspondence of the VEVs [2].

$$\langle\phi_X\rangle = \langle e^{\sigma+i\rho}\rangle,\quad \langle\phi_Y\rangle = \langle e^{-(\sigma+i\rho)}\rangle,\quad \langle\phi_Z\rangle = \langle\phi_{\tilde{Q}}\phi_Q\rangle. \tag{6.20}$$

Now, let us remind our parity conditions for component fields in vector multiplet (5.2). As one can check,

$$\sigma+i\rho \to -(\sigma+i\rho) \tag{6.21}$$

[1] The reason for taking α^{-1} not α in (6.17) is that the sign of the conserved current for $U(1)_A$ is reversed under the mirror symmetry [5].

occurs under the antipodal identification (5.1). And we choose here the parity for matter fields in SQED as

$$\phi_Q \to \phi_Q, \quad \phi_{\tilde{Q}} \to \phi_{\tilde{Q}}, \tag{6.22}$$

then, (6.20) suggests the following parity conditions for XYZ-model:

$$\phi_X \leftrightarrows \phi_Y, \quad \phi_Z \to \phi_Z. \tag{6.23}$$

The parity conditions (6.23) mean that the matter multiplets X and Y form the doublet with the parity matrix (5.6), and the matter multiplet Z is singlet under the antipodal identification.[2]

SCI of XYZ-Model

According to the formula in (5.55) and the charge assignments in (6.4), we get

$$\mathcal{I}^{\Delta}_{\text{XYZ}}(x, \alpha) = \left(x^{+\frac{2\Delta-1}{4}}\alpha^{\frac{-1}{2}}\right)\frac{(\alpha^2 x^{2(1-\Delta)}; x^4)_\infty}{(\alpha^{-2}x^{2\Delta}; x^4)_\infty} \times \frac{(\alpha^{-1}x^{(1+\Delta)}; x^2)_\infty}{(\alpha x^{(1-\Delta)}; x^2)_\infty} \tag{6.24}$$

The spatial value for $\Delta = 1/2, \alpha = 1$ provides

$$\mathcal{I}^{1/2}_{\text{XYZ}}(x, 1) = 1 + x^{1/2} + x + x^{5/2} + x^3 - x^4 + 2x^5 + x^{11/2} - x^6 - x^{13/2} + x^7 + \dots . \tag{6.25}$$

This gives totally different contributions compared with (6.15).

SCI of SQED

According to the formula (5.59) and the charge assignments in (6.11), we get

$$\begin{aligned}
&\mathcal{I}^{\Delta}_{\text{SQED}}(x, \alpha^{-1}) \\
&= \int_0^{2\pi} \frac{d\theta}{2\pi}\left(x^{+\frac{2\Delta-1}{4}}\alpha^{\frac{-1}{2}}\right)\frac{(e^{-i\theta}\alpha x^{(2-\Delta)}; x^4)_\infty}{(e^{i\theta}\alpha^{-1}x^{\Delta}; x^4)_\infty} \times \frac{(e^{i\theta}\alpha x^{(2-\Delta)}; x^4)_\infty}{(e^{-i\theta}\alpha^{-1}x^{\Delta}; x^4)_\infty} \times \frac{(x^4; x^4)_\infty}{(x^2; x^4)_\infty} \\
&+ \int_0^{2\pi} \frac{d\theta}{2\pi}\left(x^{-\frac{2\Delta-3}{4}}\alpha^{\frac{1}{2}}\right)\frac{(e^{-i\theta}\alpha x^{(4-\Delta)}; x^4)_\infty}{(e^{i\theta}\alpha^{-1}x^{(2+\Delta)}; x^4)_\infty} \times \frac{(e^{i\theta}\alpha x^{(4-\Delta)}; x^4)_\infty}{(e^{-i\theta}\alpha^{-1}x^{(2+\Delta)}; x^4)_\infty} \times \frac{(x^4; x^4)_\infty}{(x^2; x^4)_\infty}.
\end{aligned} \tag{6.26}$$

This gives

$$\mathcal{I}^{1/2}_{\text{SQED}}(x, 1) = 1 + x^{1/2} + x + x^{5/2} + x^3 - x^4 + 2x^5 + x^{11/2} - x^6 - x^{13/2} + x^7 + \dots \tag{6.27}$$

The reader can find the exact proof for the equality in Appendix C.2.

[2] An another correspondence of parity conditions is discovered in [7].

References

1. K.A. Intriligator, N. Seiberg, Mirror symmetry in three-dimensional gauge theories. Phys. Lett. **B387**, 513–519 (1996). arXiv:hep-th/9607207 [hep-th]
2. O. Aharony, A. Hanany, K.A. Intriligator, N. Seiberg, M. Strassler, Aspects of N = 2 supersymmetric gauge theories in three-dimensions. Nucl. Phys. **B499**, 67–99 (1997). arXiv:hep-th/9703110 [hep-th]
3. A. Kapustin, M.J. Strassler, On mirror symmetry in three-dimensional Abelian gauge theories. J. High Energy Phys. **9904**, 021 (1999). arXiv:hep-th/9902033 [hep-th]
4. N. Hama, K. Hosomichi, S. Lee, Notes on SUSY gauge theories on three-sphere. J. High Energy Phys. **1103**, 127 (2011). arXiv:1012.3512 [hep-th]
5. A. Kapustin, B. Willett, Generalized superconformal index for three dimensional field theories. arXiv:1106.2484 [hep-th]
6. A. Tanaka, H. Mori, T. Morita, Superconformal index on $RP^2 \times S^1$ and mirror symmetry. Phys. Rev. D **91**, 105023 (2015). arXiv:1408.3371 [hep-th]
7. A. Tanaka, H. Mori, T. Morita, Abelian 3d mirror symmetry on $RP^2 \times S^1$ with $N_f = 1$. J. High Energy Phys. **09**, 154 (2015). arXiv:1505.07539 [hep-th]

Chapter 7
Concluding Remarks

In this thesis, we performed exact calculations of the SCI based on the supersymmetric localization method. We considered supersymmetric QFT on $S^2 \times S^1_\beta$ in Chap. 4, on $RP^2 \times S^1_\beta$ in Chap. 5. By integrating out the degrees of freedom along the 2-dimensional surface, we got many-body quantum mechanics. The families of many particles coming from the reduction along the S^2 are different from the ones along the RP^2. In this sense, we may be able to regard that the difference between the SCI on $S^2 \times S^1_\beta$ and the SCI on $RP^2 \times S^1_\beta$ is the difference of the Hilbert space $\mathcal{H}$ in (3.1). And we also applied these two SCI's to check the conjectural duality, three-dimensional mirror symmetry or equivalence between XYZ-model (6.3) and SQED (6.10). As one can find in Appendix C, the equivalence can be recognized by the uses of the mathematical formulas.

$$S^2 \times S^1_\beta \text{ case : } \bullet \begin{cases} \text{Ramanujan's summation formula (C.7)} \\ \qquad\qquad + \\ \text{q-binomial formula (C.13)} \end{cases}$$

$$RP^2 \times S^1_\beta \text{ case : } \bullet \text{ q-binomial formula (C.13)}$$

Ramanujan's summation formula is necessary for summing up all contributions labelled by the monopole numbers $B \in Z$. And q-binomial formula is necessary for conducting the residue integrals, so it comes from the integral over $\theta \in [0, 2\pi]$. In later case, as one can notice, the following unnamed formulas are important.

$$(A; q)_{2l} = (A; q^2)_l (Aq; q^2)_l, \quad (A; q)_{2l+1} = (1 - A)(Aq; q^2)_l (Aq^2; q^2)_l, \quad \text{for } l \in N. \tag{7.1}$$

These formulas can be regarded as an algebraic representations of the $\pm$ holonomies along RP^2. In summary, in the context of the mirror symmetry, there are the following correspondences between algebraic mathematical formula and geometric physical object.

A. Tanaka, *Superconformal Index on* RP^2 $\times$ S^1 *and 3D Mirror Symmetry*,
Springer Theses, DOI 10.1007/978-981-10-1398-0_7

$$\text{Ramanujan's summation formula} \Leftrightarrow \text{Monopoles on } S^2, \tag{7.2}$$

$$\text{No name formulas in (7.1.1)} \Leftrightarrow \text{Holonomies along } RP^2, \tag{7.3}$$

$$\text{q-binomial formula} \Leftrightarrow \text{Holonomy along } S^1_\beta. \tag{7.4}$$

Thanks to the duality realized in such way, we can observe how the duality works in mathematically rigorous way. Realizing QFT in such way provides us rigid understandings of QFT's non-perturbative aspects, and conversely, the duality provides unexpected relationships between different mathematical objects. Therefore, the study of the dualities in quantum physics is fruitful and very interesting research area.

One more comment As noted in Chap. 5, there is different supersymmetric parity conditions. This is as follows.

$$\begin{aligned}
&A_\vartheta(\pi-\vartheta,\pi+\varphi,y) = +A_\vartheta(\vartheta,\varphi,y), \quad A_{\varphi,y}(\pi-\vartheta,\pi+\varphi,y) = -A_{\varphi,y}(\vartheta,\varphi,y),\\
&\sigma(\pi-\vartheta,\pi+\varphi,y) = +\sigma(\vartheta,\varphi,y),\\
&\lambda(\pi-\vartheta,\pi+\varphi,y) = -i\gamma_1\lambda(\vartheta,\varphi,y), \quad \overline{\lambda}(\pi-\vartheta,\pi+\varphi,y) = +i\gamma_1\overline{\lambda}(\vartheta,\varphi,y),\\
&D(\pi-\vartheta,\pi+\varphi,y) = -D(\vartheta,\varphi,y).
\end{aligned} \tag{7.5}$$

This condition also preserves SUSY and $U(1)$ Yang–Mills action (3.64). However, it breaks the invariance of the following differential operator.

$$(\partial - iA)^2, \tag{7.6}$$

because under the above transformation, we get

$$(\partial - iA)^2 \to (\partial + iA)^2. \tag{7.7}$$

In order to overcome such problem, we have to turn on two matters with $\pm$ charges respectively. We have such mattes in SQED, Q and $\tilde{Q}$, and in fact, we can observe corresponding duality [1]. In addition, one can check mirror symmetry for multi-flavored theory and inclusion of non-local operators [2].

References

1. A. Tanaka, H. Mori, T. Morita, Abelian 3d mirror symmetry on $RP^2 \times S^1$ with $N_f = 1$. J. High Energy Phys. **09**, 154 (2015). arXiv:1505.07539 [hep-th]
2. H. Mori, A. Tanaka, Varieties of Abelian mirror symmetry on $RP^2 \times S^1$. J. High Energy Phys. **02**, 088 (2016). arXiv:1512.02835 [hep-th]

Appendix A
Mathematics for the Thesis

In this appendix, we summarize and derive some mathematical formulas which are relevant in the thesis.

A.1 Trigonometric Functions

As well known, the trigonometric functions can be represented as infinite products:

$$\sin \pi z = \pi z \prod_{n=1}^{\infty} \left(1 - \frac{z^2}{n^2}\right), \quad \sinh \pi z = \pi z \prod_{n=1}^{\infty} \left(1 + \frac{z^2}{n^2}\right) \tag{A.1.1}$$

$$\cos \pi z = \prod_{n=1}^{\infty} \left(1 - \frac{z^2}{(n-\frac{1}{2})^2}\right), \quad \cosh \pi z = \prod_{n=1}^{\infty} \left(1 + \frac{z^2}{(n-\frac{1}{2})^2}\right). \tag{A.1.2}$$

One interesting application is an infinite product formula for π:

$$\begin{aligned} 1 = \sin \frac{\pi}{2} = \frac{\pi}{2} \prod_{n=1}^{\infty} \left(1 - \frac{(\frac{1}{2})^2}{n^2}\right) &= \frac{\pi}{2} \prod_{n=1}^{\infty} \left(\frac{(2n)^2 - 1}{(2n)^2}\right) \\ &= \frac{\pi}{2} \prod_{n=1}^{\infty} \left(\frac{(2n-1)(2n+1)}{(2n)^2}\right) \end{aligned} \tag{A.1.3}$$

This is called *Wallis' formula*.

A. Tanaka, *Superconformal Index on* $\mathrm{RP}^2 \times \mathrm{S}^1$ *and 3D Mirror Symmetry*,
Springer Theses, DOI 10.1007/978-981-10-1398-0

A.2 Zeta Function

We use the *zeta function regularization* throughout this thesis. This regularization corresponds to introducing a cutoff to the UV momentum [1]. The zeta function is defined by

$$\zeta(s) = \sum_{n=1}^{\infty} n^{-s} \quad \text{for } Re(s) > 1, \tag{A.2.1}$$

and it is analytically continued to whole complex plane $s \in C$. One can try to calculate particular value for fixed s by introducing UV cutoff for n. For example,

$$\begin{aligned} \zeta(0) &\sim \sum_{n=1}^{\infty} 1 \overset{0 \leftarrow \epsilon}{\longleftarrow} \sum_{n=1}^{\infty} e^{-\epsilon n} \\ &= \frac{e^{-\epsilon}}{1-e^{-\epsilon}} = \frac{1}{e^{\epsilon}-1} = \frac{1}{\epsilon\left(1+\frac{1}{2}\epsilon+\mathcal{O}(\epsilon^2)\right)} \\ &= \frac{1}{\epsilon}\left(1-\frac{1}{2}\epsilon+\mathcal{O}(\epsilon^2)\right) = \frac{1}{\epsilon}-\frac{1}{2}+\mathcal{O}(\epsilon), \end{aligned} \tag{A.2.2}$$

in this regularization, the "scale" for the cutoff corresponds to ϵ and UV limit is $\epsilon \to 0$. Obviously, the divergent first term in (A.2.2) represents UV divergence. Now we take the following renormalization:

$$\zeta(0) \sim \lim_{\epsilon \to 0}\left[\sum_{n=1}^{\infty} e^{-\epsilon n} - \frac{1}{\epsilon}\right] = -\frac{1}{2}. \tag{A.2.3}$$

In fact, it is known that this procedure reproduces the precise analytic continued value for $\zeta(0)$.

The 1st derivative of zeta function We would like to derive the value for $\zeta'(0)$ here. By differentiating (A.2.1) with s, we can get

$$\zeta'(s) = -\sum_{n=1}^{\infty} n^{-s} \log n. \tag{A.2.4}$$

So the value for $s = 0$ may be

$$\zeta'(0) \sim -\sum_{n=1}^{\infty} \log n = -\log \prod_{n=1}^{\infty} n. \tag{A.2.5}$$

This divergence can be regularize by using Wallis' formula and the regularized value of $\zeta(0)$ as follows. 1st, by deforming Wallis' formula,

$$
\begin{aligned}
\frac{\pi}{2} &= \prod_{n=1}^{\infty} \frac{(2n)^2}{(2n-1)(2n+1)} = \prod_{n=1}^{\infty} \frac{(2n)^4}{(2n)^2(2n-1)(2n+1)} \\
&= \left(\prod_{n=1}^{\infty} 2^4\right)\left(\prod_{n=1}^{\infty} n\right)^4 \left(\prod_{n=1}^{\infty} \frac{1}{(2n)(2n-1)}\right)\left(\prod_{n=1}^{\infty} \frac{1}{(2n)(2n+1)}\right) \\
&= \left(2^{4\sum_{n=1}^{\infty} 1}\right)\left(\prod_{n-1}^{\infty} n\right)^4 \left(\prod_{n-1}^{\infty} \frac{1}{n}\right)\left(\prod_{n-1}^{\infty} \frac{1}{n}\right) \\
&\sim \left(2^{4\zeta(0)}\right)\left(\prod_{n=1}^{\infty} n\right)^2 = \left(2^{-2}\right)\left(\prod_{n=1}^{\infty} n\right)^2,
\end{aligned}
\tag{A.2.6}
$$

2nd, by taking $\sqrt{\ }$, we arrive at

$$
\prod_{n=1}^{\infty} n \sim \sqrt{2\pi}. \tag{A.2.7}
$$

Then, by substituting it to (A.2.5), we get

$$
\zeta'(0) = -\frac{1}{2} \log 2\pi. \tag{A.2.8}
$$

This derivation is slightly dangerous but it is found in [2], and it gives correct answer.

A.3 Gaussian Integrals

The gaussian integral

$$
\int_{-\infty}^{\infty} dx\, e^{-\frac{1}{2}x^2} = \sqrt{2\pi} \tag{A.3.1}
$$

is the most important integral in this thesis. Here, we summarize basic facts of Gaussian integrals of bosonic degrees of freedom x_i and fermonic degrees of free dom ψ_i.

Bosonic Case

$$
\text{Real Gaussian}: \int \prod_i \frac{dx_i}{\sqrt{2\pi}} e^{-\frac{1}{2}\sum_{ij} x_i M_{ij} x_j} = \frac{1}{\sqrt{\det M_{ij}}} \tag{A.3.2}
$$

$$\text{Complex Gaussian}: \int \prod_i \frac{dz_i d\bar{z}_i}{2\pi} e^{-\frac{1}{2}\sum_{ij} \bar{z}_i M_{ij} z_j} = \frac{1}{\det M_{ij}} \tag{A.3.3}$$

Fermionic Case

$$\text{Real Gaussian}: \int \prod_i d\psi_i e^{-\frac{1}{2}\sum_{ij} \psi_i M_{ij} \psi_j} = \sqrt{\det M_{ij}} \tag{A.3.4}$$

$$\text{Complex Gaussian}: \int \prod_i d\psi_i d\bar{\psi}_i e^{-\frac{1}{2}\sum_{ij} \bar{\psi}_i M_{ij} \psi_j} = \det M_{ij} \tag{A.3.5}$$

References

1. J. Polchinski, String Theory. An Introduction to the Bosonic String, vol. 1 (Cambridge University Press, Cambridge, 2007).
2. J. Sondow, Analytic continuation of Riemann's zeta function and values at negative integers via Euler's transformation of series. Proc. Am. Math. Soc. **120**, 421–424 (1994).

Appendix B
Monopole Spherical Harmonics

As well known in the context of Schödinger equation for spherically symmetric system, the spherical harmonics $Y_{jm}(\vartheta, \varphi)$ diagonalizes the Laplacian on S^2:

$$\nabla_i \nabla^i Y_{jm}(\vartheta, \varphi) = \left(\frac{1}{\sin \vartheta} \partial_\vartheta \sin \vartheta \partial_\vartheta + \frac{1}{\sin^2 \vartheta} \partial_\varphi^2 \right) Y_{jm}(\vartheta, \varphi)$$
$$= -j(j+1) Y_{jm}(\vartheta, \varphi). \tag{B.0.1}$$

This is a consequence of the fact that the Laplacian $\nabla_i \nabla^i$ on S^2 can be regarded as the squared orbital angular momentum $\vec{L}^2$. Here, let us remind the definition for the orbital angular momentum operators:

$$L_1 \pm i L_2 = e^{i\varphi} \Big(\pm \partial_\vartheta + i \cot \vartheta \partial_\varphi \Big), \quad L_3 = -i \partial_\varphi. \tag{B.0.2}$$

L_1, L_2, L_3 satisfy the $SU(2)$ algebra:

$$[L_A, L_B] = i \epsilon_{ABC} L_C. \tag{B.0.3}$$

The spectrum of $-\nabla_i \nabla^i = \vec{L}^2 = L_1^2 + L_2^2 + L_3^2$ is purely determined by this $SU(2)$ algebraic structure:

$$\vec{L}^2 Y_{jm}(\vartheta, \varphi) = j(j+1) Y_{jm}(\vartheta, \varphi), \tag{B.0.4}$$
$$L_3 Y_{jm}(\vartheta, \varphi) = m Y_{jm}(\vartheta, \varphi). \tag{B.0.5}$$

In this appendix, we review generalizations of this construction.

Monopole background Consider a background $U(1)$ gauge field

$$A_{\text{mon}} = \frac{B}{2}(\kappa - \cos \vartheta) d\varphi, \tag{B.0.6}$$

A. Tanaka, *Superconformal Index on* RP^2 × S^1 *and 3D Mirror Symmetry*,
Springer Theses, DOI 10.1007/978-981-10-1398-0

where κ is $+1$ when we take a coordinate patch around north pole; $0 \leq \vartheta < \pi$, and -1 when we take a coordinate patch around south pole; $0 < \vartheta \leq \pi$. The gauge field around north pole, say A^n, and the gauge field around south pole, say A^s are related by the following gauge transformation:

$$A^n_{\text{mon}} = A^s_{\text{mon}} + ig^{-1}dg, \quad g = e^{Bi\varphi}. \tag{B.0.7}$$

Now, in order to define the gauge transformation g as single valued function on S^2, we have to take $B \in Z$. This is famous Dirac's quantization condition for monopole charge.

Monopole harmonics By using the background gauge field (B.0.6), we can generalize the orbital angular momentum operators (B.0.2):

$$J_1 \pm iJ_2 = e^{i\varphi}\left(\pm\partial_\vartheta + i\cot\vartheta(\partial_\varphi - iA_\varphi) + \frac{B}{2}\sin\vartheta\right), \quad J_3 = -i\partial_\varphi \mp \frac{B}{2}. \tag{B.0.8}$$

One may wonder the physical meaning of this definition. It becomes clear when we represent them by using $x_1 = r\sin\vartheta\cos\varphi$, $x_2 = r\sin\vartheta\sin\varphi$, $x_3 = r\cos\vartheta$:

$$\vec{J} = \vec{r} \times \left(-i\vec{\nabla} + \vec{A}_{\text{mon}}\right) + \frac{B}{2}\frac{\vec{r}}{r}. \tag{B.0.9}$$

$\vec{J}$ is composed of orbital angular momentum under the background gauge field (B.0.6) and the angular moment of the monopole itself. Note that the value for A_φ on north pole patch and south pole patch are different, so $J_1 \pm J_2$ are not usual differential operators. Precisely speaking, the operators (B.0.8) act on not functions but sections of certain non-trivial vector bundle. These operators satisfy

$$[J_A, J_B] = i\epsilon_{ABC}J_C. \tag{B.0.10}$$

In the following sections, we briefly summarize the eigenstates for $\vec{J}^2$, J_3:

$$\vec{J}^2|j,m\rangle = j(j+1)|j,m\rangle, \tag{B.0.11}$$
$$J_3|j,m\rangle = m|j,m\rangle, \tag{B.0.12}$$

with spin 0, 1/2, 1, respectively. For later use, we define monopole covariant derivative

$$\mathfrak{D}_i := \nabla_i - iA^{\text{mon}}{}_i, \tag{B.0.13}$$

where ∇_i is defined in (3.24), the usual covariant derivative with respect to the spin connection.

B.1 Scalar Harmonics $Y_{\frac{B}{2},jm}$

With a spin 0 scalar field, one can verify

$$\mathfrak{D}_i\mathfrak{D}^i = -\left(\vec{J}^2 - \frac{B^2}{2^2}\right). \tag{B.1.1}$$

This fact means that we can diagonalize the monopole Laplacian $\mathfrak{D}_i\mathfrak{D}^i$ on S^2 with the state satisfying (B.0.11) and (B.0.12). Let us define the spin zero wave function as $Y_{\frac{B}{2},jm}(\vartheta,\varphi)$, then we get

$$\mathfrak{D}_i\mathfrak{D}^i Y_{\frac{B}{2},jm}(\vartheta,\varphi) = -\left(j(j+1) - \frac{B^2}{2^2}\right) Y_{\frac{B}{2},jm}(\vartheta,\varphi). \tag{B.1.2}$$

By using well known argument of orthogonality based on integration by parts, we can also derive

$$\int \sin\vartheta d\vartheta d\varphi\, Y^*_{\frac{B}{2},jm}(\vartheta,\varphi) Y_{\frac{B}{2},j'm'}(\vartheta,\varphi) = \delta_{jj'}\delta_{mm'}. \tag{B.1.3}$$

If and only if $j \geq |\frac{B}{2}|$, $Y_{\frac{B}{2},jm}$ becomes normalizable. See [1] for more details.

B.2 Spinor Harmonics $\Upsilon_{\frac{B}{2},jm}$, $\Psi_{\frac{B}{2},jm}$

Spin 1/2 monopole angular momentum operators satisfy the following relation.

$$\vec{J}^{\,2}_{spinor} = -(\gamma^i\mathfrak{D}_i)^2 - \frac{1}{4} + \left(\frac{B}{2}\right)^2. \tag{B.2.1}$$

Therefore, by taking square root of this eigenvalues, we can diagonalize the monopole Dirac operator $-i\gamma^i\mathfrak{D}_i$ on S^2 with the spin 1/2 state satisfying (B.0.11) and (B.0.12).

Eigenspinors for $-i\gamma^i\mathfrak{D}_i$

As one can notice, there must be two modes:

$$-i\gamma^i\mathfrak{D}_i\Upsilon^{\pm}_{\frac{B}{2},jm}(\vartheta,\varphi) = \pm\mu_{j\frac{B}{2}}\Upsilon^{\pm}_{\frac{B}{2},jm}(\vartheta,\varphi), \quad \mu_{j\frac{B}{2}} = \frac{\sqrt{(2j+1)^2 - B^2}}{2} \tag{B.2.2}$$

where the two modes are exchanged by the multiplication of γ_3:

$$\gamma_3\Upsilon^{\pm}_{\frac{B}{2},jm}(\vartheta,\varphi) = \Upsilon^{\mp}_{\frac{B}{2},jm}(\vartheta,\varphi). \tag{B.2.3}$$

And the normalizability requires $j \geq \frac{|B|}{2} - \frac{1}{2}$. When $j = \frac{|B|}{2} - \frac{1}{2}$, we have one zero mode:

$$- i\gamma^i \mathfrak{D}_i \Upsilon^0_{\frac{B}{2},jm}(\vartheta, \varphi) = 0, \tag{B.2.4}$$

$$\gamma_3 \Upsilon^0_{\frac{B}{2},jm}(\vartheta, \varphi) = \text{sign}(B)\Upsilon^0_{\frac{B}{2},jm}(\vartheta, \varphi). \tag{B.2.5}$$

$\Psi^{\epsilon=\pm,0}_{\frac{B}{2},jm}(\vartheta, \varphi)$ are orthonormal:

$$\int \sin\vartheta d\vartheta d\varphi\ \Upsilon^{\epsilon}_{\frac{B}{2},jm}(\vartheta, \varphi)^\dagger \Upsilon^{\epsilon'}_{\frac{B}{2},j'm'}(\vartheta, \varphi) = \delta^{\epsilon\epsilon'}\delta_{jj'}\delta_{mm'}. \tag{B.2.6}$$

See the appendix of [2] for more details.

Eigenspinors for $-i\gamma_3\gamma^i\mathfrak{D}_i$

One can construct eigenspjnors for $-i\gamma_3\gamma^i\mathfrak{D}_i$ by taking

$$\Psi^{\pm}_{\frac{B}{2},jm}(\vartheta, \varphi) = (1 - i\gamma_3)\Upsilon^{\pm}_{\frac{B}{2},jm}(\vartheta, \varphi). \tag{B.2.7}$$

These spinors give following formula

$$-i\gamma_3\gamma^i\mathfrak{D}_i\Psi^{\pm}_{\frac{B}{2},jm}(\vartheta, \varphi) = \pm i\mu_{j\frac{B}{2}}\Psi^{\pm}_{\frac{B}{2},jm}(\vartheta, \varphi) \tag{B.2.8}$$

We define corresponding $\overline{\Psi}$ as

$$\int \sin\vartheta d\vartheta d\varphi\ \overline{\Psi}^{\epsilon}_{\frac{B}{2},jm}(\vartheta, \varphi)\gamma_3\Psi^{\epsilon'}_{\frac{B}{2},j'm'}(\vartheta, \varphi) = \delta^{\epsilon\epsilon'}\delta_{jj'}\delta_{mm'}. \tag{B.2.9}$$

B.3 Vector Harmonics $V^i_{\frac{B}{2},jm}$

By repeating procedure similar to the case represented above, we can make vector harmonics [3]. However it is somewhat complicated, so we would like to concentrate on the case of

$$B = 0, \quad \nabla_i V^i_{jm}(\vartheta, \varphi) = 0. \tag{B.3.1}$$

This vector satisfies the following formulas [4]:

$$\nabla_1 V^2_{jm}(\vartheta, \varphi) - \nabla_2 V^1_{jm}(\vartheta, \varphi) = \sqrt{j(j+1)}Y_{jm}(\vartheta, \varphi), \quad (\text{for } j \geq 1) \tag{B.3.2}$$

$$\nabla_1 V^2_{jm}(\vartheta, \varphi) - \nabla_2 V^1_{jm}(\vartheta, \varphi) = 0, \quad (\text{for } j = -1). \tag{B.3.3}$$

When $j = \frac{|B|}{2}$, the mode with (B.3.1) becomes zero. Orthonormality condition is

$$\int \sin\vartheta d\vartheta d\varphi\ V^{i}_{jm}(\vartheta,\varphi)V^{i}_{j'm'}(\vartheta,\varphi) = \delta_{jj'}\delta_{mm'}. \tag{B.3.4}$$

References

1. T.T. Wu, C.N. Yang, Dirac monopole without strings: monopole harmonics. Nucl. Phys. **B107**, 365 (1976).
2. J. Lindkvist, R-symmetry charges of monopole operators (2010).
3. E.J. Weinberg, Monopole vector spherical harmonics. Phys. Rev. **D49**, 1086–1092 (1994). arXiv:hep-th/9308054 [hep-th].
4. D. Gang, Chern-Simons theory on L(p, q) lens spaces and Localization. arXiv: 0912.4664 [hep-th].

Appendix C
Proof of $\mathcal{I}^{\Delta}_{XYZ} = \mathcal{I}^{\Delta}_{SQED}$

C.1 $M^2 = S^2$ Case

The following argument is originally found by [1]. In order to calculate this complex integral (6.17), it is useful to change the integration variable from θ to $z = e^{i\theta}$:

$$(6.17) = \sum_{B\in Z}\oint \frac{dz}{2\pi i z}\left(x^{(1-\Delta)}\alpha\right)^{|B|}\frac{(z^{-1}\alpha x^{2-\Delta+|B|};x^2)_\infty}{(z\alpha^{-1}x^{\Delta+|B|};x^2)_\infty}\times\frac{(z\alpha x^{2-\Delta+|B|};x^2)_\infty}{(z^{-1}\alpha^{-1}x^{\Delta+|B|};x^2)_\infty}, \tag{C.1.1}$$

then, the problem is which poles are chosen. We assume here that

$$|\alpha^{-1}x^{\Delta+|B|}| < 1. \tag{C.1.2}$$

Then, the relevant residues are located at

$$z_l = x^{2l+\Delta+|B|}\alpha^{-1}, \quad l = 0, 1, 2, \ldots \tag{C.1.3}$$

and the integral becomes

$$(C.1.1) = \sum_{B\in Z}\sum_{l=0}^{\infty}\left(x^{(1-\Delta)}\alpha\right)^{|B|}\frac{(\alpha^2 x^{-2(l-1+\Delta)};x^2)_\infty}{(\alpha^{-2}x^{2(l+\Delta+|B|)};x^2)_\infty} \times\frac{(x^{2(1+l)+2|B|};x^2)_\infty}{(x^2;x^2)_\infty}\times\frac{1}{(x^{-2l};x^2)_l}, \tag{C.1.4}$$

where $(A;q)_l = \prod_{n=0}^{l-1}(1-Aq^n)$. Now, we can observe the following fact: the $|B|$ in the series (C.1.4) can be replaced by B [1, 2], and the following formula:

A. Tanaka, *Superconformal Index on* $\mathrm{RP}^2 \times \mathrm{S}^1$ *and 3D Mirror Symmetry*,
Springer Theses, DOI 10.1007/978-981-10-1398-0

$$(Ax^{2B}; x^2)_\infty = \frac{(A; x^2)_\infty}{(A; x^2)_B}, \tag{C.1.5}$$

where $(A; q)_{-l} = \prod_{n=1}^{l}(1 - Aq^{-n})^{-1}$ for $l > 0$. Then,

$$\begin{aligned}(\text{C.1.4}) = &\sum_{l=0}^{\infty} \frac{(\alpha^2 x^{-2(l-1+\Delta)}; x^2)_\infty}{(\alpha^{-2}x^{2(l+\Delta)}; x^2)_\infty} \frac{(x^{2(1+l)}; x^2)_\infty}{(x^2; x^2)_\infty} \frac{1}{(x^{-2l}; x^2)_l} \\ &\times \underbrace{\sum_{B\in Z} \left(x^{(1-\Delta)}\alpha\right)^B \frac{(\alpha^{-2}x^{2(l+\Delta)}; x^2)_B}{(x^{2(1+l)}; x^2)_B}}_{\boxed{\text{1st key terms}}},\end{aligned} \tag{C.1.6}$$

Now, we use the following formula in order to deform the $\boxed{\text{1st key terms}}$:

Ramanujan's summation formula [3]

$$\sum_{B\in Z} z^B \frac{(a, q)_B}{(b, q)_B} = \frac{(q; q)_\infty (\frac{b}{a}; q)_\infty (az; q)_\infty (\frac{q}{az}; q)_\infty}{(b; q)_\infty (\frac{q}{a}; q)_\infty (z; q)_\infty (\frac{b}{az}; q)_\infty} \tag{C.1.7}$$

In our case (C.1.6),

$$q = x^2, \quad z = \left(x^{(1-\Delta)}\alpha\right), \quad a = (\alpha^{-2}x^{2(l+\Delta)}), \quad b = x^{2(1+l)}. \tag{C.1.8}$$

Then,

$$\boxed{\text{1st key terms}} = \frac{(x^2; x^2)_\infty (\alpha^2 x^{2(1-\Delta)}; x^2)_\infty (\alpha^{-1}x^{2l+\Delta+1}; x^2)_\infty (\alpha x^{1-\Delta-2l}; x^2)_\infty}{(x^{2(1+l)}; x^2)_\infty (\alpha^2 x^{-2(l+\Delta-1)}; x^2)_\infty (\alpha x^{1-\Delta}; x^2)_\infty (\alpha x^{1-\Delta}; x^2)_\infty}. \tag{C.1.9}$$

By substituting it into (C.1.6), we get

$$\begin{aligned}&(\text{C.1.6}) \\ &= \sum_{l=0}^{\infty} \frac{(\alpha^2 x^{-2(l-1+\Delta)}; x^2)_\infty}{(\alpha^{-2}x^{2(l+\Delta)}; x^2)_\infty} \frac{(x^{2(1+l)}; x^2)_\infty}{(x^2; x^2)_\infty} \frac{1}{(x^{-2l}; x^2)_l} \\ &\quad \times \frac{(x^2; x^2)_\infty (\alpha^2 x^{2(1-\Delta)}; x^2)_\infty (\alpha^{-1}x^{2l+\Delta+1}; x^2)_\infty (\alpha x^{1-\Delta-2l}; x^2)_\infty}{(x^{2(1+l)}; x^2)_\infty (\alpha^2 x^{-2(l+\Delta-1)}; x^2)_\infty (\alpha x^{1-\Delta}; x^2)_\infty (\alpha x^{1-\Delta}; x^2)_\infty} \\ &= \sum_{l=0}^{\infty} \frac{(-1)^l x^{-l(l+1)} \left(\alpha^2 x^{-2(-1+\Delta)}\right)^l (\alpha^{-2}x^{2(-1+\Delta)}x^2; x^2)_l (\alpha^2 x^{-2(-1+\Delta)}; x^2)_\infty}{(\alpha^{-2}x^{2\Delta}; x^2)_\infty / (\alpha^{-2}x^{2\Delta}; x^2)_l} \\ &\quad \times \frac{1}{(-1)^l x^{-l(l+1)} (x^2; x^2)_l}\end{aligned}$$

$$
\begin{aligned}
& (\alpha^2 x^{2(1-\Delta)}; x^2)_\infty \frac{(\alpha^{-1} x^{\Delta+1}; x^2)_\infty}{(\alpha^{-1} x^{\Delta+1}; x^2)_l} (-1)^l x^{-l(l+1)} \left(\alpha x^{1-\Delta}\right)^l \\
& \times \frac{(\alpha^{-1} x^{\Delta-1} x^2; x^2)_l (\alpha x^{1-\Delta}; x^2)_\infty}{(-1)^l x^{-l(l+1)} \left(\alpha^2 x^{-2(\Delta-1)}\right)^l (\alpha^{-2} x^{2(\Delta-1)} x^2; x^2)_l} \\
& \qquad (\alpha^2 x^{-2(\Delta-1)}; x^2)_\infty (\alpha x^{1-\Delta}; x^2)_\infty (\alpha x^{1-\Delta}; x^2)_\infty \\
= & \frac{(\alpha^2 x^{-2(-1+\Delta)}; x^2)_\infty (\alpha^{-1} x^{\Delta+1}; x^2)_\infty}{(\alpha^{-2} x^{2\Delta}; x^2)_\infty (\alpha x^{1-\Delta}; x^2)_\infty} \underbrace{\sum_{l=0}^{\infty} \frac{(\alpha^{-2} x^{2\Delta}; x^2)_l}{(x^2; x^2)_l} \left(\alpha x^{1-\Delta}\right)^l}_{\boxed{\text{2nd key terms}}} \qquad \text{(C.1.10)}
\end{aligned}
$$

Here, we used the following formulas.

$$
(x^{-2l}; x^2)_l = (-1)^l x^{-2l(l+1)} (x^2; x^2)_l, \qquad \text{(C.1.11)}
$$

$$
(A x^{-2l}; x^2)_\infty = (-1)^l x^{-2l(l+1)} A^l (A^{-1} x^2; x^2)_l (A; x^2)_\infty \qquad \text{(C.1.12)}
$$

The final key is the following formula:

q-binomial formula [3]

$$
\sum_{l=0}^{\infty} \frac{(A; q)_l}{(q; q)_l} Z^l = \frac{(AZ; q)_\infty}{(Z; q)_\infty} \qquad \text{(C.1.13)}
$$

In our case, (C.1.10),

$$
q = x^2, \quad A = \alpha^{-2} x^{2\Delta}, \quad Z = \alpha x^{1-\Delta}, \qquad \text{(C.1.14)}
$$

so we get

$$
\boxed{\text{2nd key terms}} = \frac{(\alpha^{-1} x^{1+\Delta}; x^2)_\infty}{(\alpha x^{1-\Delta}; x^2)_\infty} \qquad \text{(C.1.15)}
$$

Substituting it into (C.1.10), we finally arrived at

$$
\begin{aligned}
\text{(C.1.10)} &= \frac{(\alpha^2 x^{-2(-1+\Delta)}; x^2)_\infty (\alpha^{-1} x^{\Delta+1}; x^2)_\infty}{(\alpha^{-2} x^{2\Delta}; x^2)_\infty (\alpha x^{1-\Delta}; x^2)_\infty} \frac{(\alpha^{-1} x^{1+\Delta}; x^2)_\infty}{(\alpha x^{1-\Delta}; x^2)_\infty} \\
&= \left(\frac{(\alpha^{-1} x^{1+\Delta}; x^2)_\infty}{(\alpha x^{1-\Delta}; x^2)_\infty} \right)^2 \frac{(\alpha^2 x^{2(1-\Delta)}; x^2)_\infty}{(\alpha^{-2} x^{2\Delta}; x^2)_\infty}. \qquad \text{(C.1.16)}
\end{aligned}
$$

This is exactly identical to the SCI of XYZ-model (6.14).

C.2 $M^2 = RP^2$ Case

The following argument is based on our original work [4]. In order to calculate this complex integral (6.26), it is useful to change the integration variable from θ to $z,\ w = e^{i\theta}$:

$$
\begin{aligned}
(6.26) = &\oint \frac{dz}{2\pi i z}\left(x^{+\frac{2\Delta-1}{4}}\alpha^{\frac{-1}{2}}\right)\frac{(z^{-1}\alpha x^{(2-\Delta)};x^4)_\infty}{(z\alpha^{-1}x^{\Delta};x^4)_\infty}\times\frac{(z\alpha x^{(2-\Delta)};x^4)_\infty}{(z^{-1}\alpha^{-1}x^{\Delta};x^4)_\infty}\\
&\times\frac{(x^4;x^4)_\infty}{(x^2;x^4)_\infty}+\oint\frac{dw}{2\pi i z}\left(x^{-\frac{2\Delta-3}{4}}\alpha^{\frac{1}{2}}\right)\frac{(w^{-1}\alpha x^{(4-\Delta)};x^4)_\infty}{(w\alpha^{-1}x^{(2+\Delta)};x^4)_\infty}\\
&\times\frac{(w\alpha x^{(4-\Delta)};x^4)_\infty}{(w^{-1}\alpha^{-1}x^{(2+\Delta)};x^4)_\infty}\times\frac{(x^4;x^4)_\infty}{(x^2;x^4)_\infty}.
\end{aligned}
\tag{C.2.1}
$$

We take same assumption (C.1.2):

$$
|\alpha^{-1}x^{\Delta+|B|}|<1,\quad B=0,2. \tag{C.2.2}
$$

Then, the relevant residues are

$$
z_l=\alpha^{-1}x^{\Delta+4l},\quad l=0,1,2,\dots\quad \text{for upper integral in (C.2.1)}, \tag{C.2.3}
$$

$$
w_l=\alpha^{-1}x^{2+\Delta+4l},\quad l=0,1,2,\dots\quad \text{for lower integral in (C.2.1)}. \tag{C.2.4}
$$

The residue integral becomes

$$
\begin{aligned}
&(\text{C.2.1})\\
&=\sum_{l=0}^{\infty}\left(x^{+\frac{2\Delta-1}{4}}\alpha^{\frac{-1}{2}}\right)\frac{(\alpha^2x^{(2-2\Delta-4l)};x^4)_\infty}{(\alpha^{-2}x^{2\Delta+4l};x^4)_\infty}\times\frac{(x^{(2+4l)};x^4)_\infty}{(x^4;x^4)_\infty(x^{-4l};x^4)_l}\times\frac{(x^4;x^4)_\infty}{(x^2;x^4)_\infty}\\
&\quad+\sum_{l=0}^{\infty}\left(x^{-\frac{2\Delta-3}{4}}\alpha^{\frac{1}{2}}\right)\frac{(\alpha^2x^{(2-2\Delta-4l)};x^4)_\infty}{(\alpha^{-2}x^{(4+2\Delta+4l)};x^4)_\infty}\times\frac{(x^{(6+4l)};x^4)_\infty}{(x^4;x^4)_\infty(x^{-4l};x^4)_l}\times\frac{(x^4;x^4)_\infty}{(x^2;x^4)_\infty}\\
&=\sum_{l=0}^{\infty}\left(x^{+\frac{2\Delta-1}{4}}\alpha^{\frac{-1}{2}}\right)\frac{(-1)^lx^{-4l(l+1)}\left(\alpha^2x^{(2-2\Delta)}\right)^l(\alpha^{-2}x^{-(2-2\Delta)}x^4;x^4)_l(\alpha^2x^{(2-2\Delta)};x^4)_\infty}{(\alpha^{-2}x^{2\Delta};x^4)_\infty/(\alpha^{-2}x^{2\Delta};x^4)_l}\\
&\quad\times\frac{(x^2;x^4)_\infty/(x^2;x^4)_l}{(x^4;x^4)_\infty(-1)^lx^{-4l(l+1)}(x^4;x^4)_l}\times\frac{(x^4;x^4)_\infty}{(x^2;x^4)_\infty}\\
&\quad+\sum_{l=0}^{\infty}\left(x^{-\frac{2\Delta-3}{4}}\alpha^{\frac{1}{2}}\right)\frac{(-1)^lx^{-4l(l+1)}\left(\alpha^2x^{(2-2\Delta)}\right)^l(\alpha^{-2}x^{-(2-2\Delta)}x^4;x^4)_l(\alpha^2x^{(2-2\Delta)};x^4)_\infty}{(\alpha^{-2}x^{(4+2\Delta)};x^4)_\infty/(\alpha^{-2}x^{(4+2\Delta)};x^4)_l}\\
&\quad\times\frac{(x^6;x^4)_\infty/(x^6;x^4)_l}{(x^4;x^4)_\infty(-1)^lx^{-4l(l+1)}(x^4;x^4)_l}\times\frac{(x^4;x^4)_\infty}{(x^2;x^4)_\infty}\\
&=\left(x^{+\frac{2\Delta-1}{4}}\alpha^{\frac{-1}{2}}\right)\frac{(\alpha^2x^{(2-2\Delta)};x^4)_\infty}{(\alpha^{-2}x^{2\Delta};x^4)_\infty}\sum_{l=0}^{\infty}\frac{(\alpha^{-2}x^{(2\Delta+2)};x^4)_l(\alpha^{-2}x^{2\Delta};x^4)_l}{(x^2;x^4)_l(x^4;x^4)_l}\left(\alpha^2x^{(2-2\Delta)}\right)^l
\end{aligned}
$$

$$+\left(x^{-\frac{2\Delta-3}{4}}\alpha^{\frac{1}{2}}\right)\frac{(\alpha^2 x^{(2-2\Delta)};x^4)_\infty}{(\alpha^{-2}x^{(4+2\Delta)};x^4)_\infty}\frac{(x^6;x^4)_\infty}{(x^2;x^4)_\infty}\sum_{l=0}^{\infty}\frac{(\alpha^{-2}x^{(2+2\Delta)};x^4)_l(\alpha^{-2}x^{(4+2\Delta)};x^4)_l}{(x^4;x^4)_l(x^6;x^4)_l}$$
$$\times\left(\alpha^2 x^{(2-2\Delta)}\right)^l, \tag{C.2.5}$$

where we used the following formulas.

$$(x^{-4l};x^4)_l = (-1)^l x^{-4l(l+1)}(x^4;x^4)_l, \tag{C.2.6}$$
$$(Ax^{-4l};x^2)_\infty = (\ \ 1)^l x^{-4l(l+1)}A^l(A^{-1}x^4;x^4)_l(A;x^4)_\infty. \tag{C.2.7}$$

We can deform the pre factor of lower term in (C.2.5) as follows:

$$\left(x^{-\frac{2\Delta-3}{4}}\alpha^{\frac{1}{2}}\right)\frac{(\alpha^2 x^{(2-2\Delta)};x^4)_\infty}{(\alpha^{-2}x^{(4+2\Delta)};x^4)_\infty}\frac{(x^6;x^4)_\infty}{(x^2;x^4)_\infty}$$
$$=\left(x^{+\frac{2\Delta-1}{4}}\alpha^{\frac{-1}{2}}\right)x^{-\frac{4\Delta-4}{4}}\alpha^1\frac{(\alpha^2 x^{(2-2\Delta)};x^4)_\infty}{(\alpha^{-2}x^{2\Delta};x^4)_\infty}\frac{(\alpha^{-2}x^{2\Delta};x^4)_\infty}{(\alpha^{-2}x^{(4+2\Delta)};x^4)_\infty}\frac{(x^6;x^4)_\infty}{(x^2;x^4)_\infty}$$
$$=\left(x^{+\frac{2\Delta-1}{4}}\alpha^{\frac{-1}{2}}\right)x^{1-\Delta}\alpha^1\frac{(\alpha^2 x^{(2-2\Delta)};x^4)_\infty}{(\alpha^{-2}x^{2\Delta};x^4)_\infty}\frac{(1-\alpha^{-2}x^{2\Delta})}{1-x^2}. \tag{C.2.8}$$

Then

(C.2.5)

$$=\left(x^{+\frac{2\Delta-1}{4}}\alpha^{\frac{-1}{2}}\right)\frac{(\alpha^2 x^{(2-2\Delta)};x^4)_\infty}{(\alpha^{-2}x^{2\Delta};x^4)_\infty}\sum_{l=0}^{\infty}\left[\frac{(\alpha^{-2}x^{(2\Delta+2)};x^4)_l(\alpha^{-2}x^{2\Delta};x^4)_l}{(x^2;x^4)_l(x^4;x^4)_l}\left(\alpha^2x^{(2-2\Delta)}\right)^l\right.$$
$$\left.+\frac{(1-\alpha^{-2}x^{2\Delta})}{1-x^2}\frac{(\alpha^{-2}x^{(2+2\Delta)};x^4)_l(\alpha^{-2}x^{(4+2\Delta)};x^4)_l}{(x^4;x^4)_l(x^6;x^4)_l}\left(\alpha^2x^{(2-2\Delta)}\right)^l\right]$$
$$=\left(x^{+\frac{2\Delta-1}{4}}\alpha^{\frac{-1}{2}}\right)\frac{(\alpha^2 x^{(2-2\Delta)};x^4)_\infty}{(\alpha^{-2}x^{2\Delta};x^4)_\infty}\sum_{l=0}^{\infty}\left[\frac{(\alpha^{-2}x^{2\Delta};x^2)_{2l}}{(x^2;x^2)_{2l}}\left(\alpha x^{(1-\Delta)}\right)^{2l}\right.$$
$$\left.+\frac{(\alpha^{-2}x^{2\Delta};x^2)_{2l+1}}{(x^2;x^2)_{2l+1}}\left(\alpha x^{(1-\Delta)}\right)^{2l+1}\right]$$
$$=\left(x^{+\frac{2\Delta-1}{4}}\alpha^{\frac{-1}{2}}\right)\frac{(\alpha^2 x^{(2-2\Delta)};x^4)_\infty}{(\alpha^{-2}x^{2\Delta};x^4)_\infty}\sum_{k=0}^{\infty}\left[\frac{(\alpha^{-2}x^{2\Delta};x^2)_k}{(x^2;x^2)_k}\left(\alpha x^{(1-\Delta)}\right)^k\right]. \tag{C.2.9}$$

Here we used

$$(A;q)_{2l} = (A;q^2)_l(Aq;q^2)_l, \quad (A;q)_{2l+1} = (1-A)(Aq;q^2)_l(Aq^2;q^2)_l. \tag{C.2.10}$$

Now, we can use the q-binomial formula (C.1.13) :

$$\sum_{k=0}^{\infty} \frac{(A;q)_k}{(q;q)_k} Z^k = \frac{(AZ;q)_\infty}{(Z;q)_\infty}, \tag{C.2.11}$$

then we arrive at

$$\text{(C.2.9)} = \left(x^{+\frac{2\Delta-1}{4}} \alpha^{\frac{-1}{2}}\right) \frac{(\alpha^2 x^{(2-2\Delta)}; x^4)_\infty}{(\alpha^{-2} x^{2\Delta}; x^4)_\infty} \frac{(\alpha^{-1} x^{1+\Delta}; x^2)_\infty}{(\alpha x^{1-\Delta}; x^2)_\infty}. \tag{C.2.12}$$

This is exactly (6.24).

References

1. C. Krattenthaler, V. Spiridonov, G. Vartanov, Superconformal indices of three-dimensional theories related by mirror symmetry. J. High Energy Phys. **1106**, 008 (2011). arXiv:1103.4075 [hep-th].
2. A. Kapustin, B. Willett, Generalized superconformal index for three dimensional field theories. arXiv:1106.2484 [hep-th].
3. G. Gasper, M. Rahman, Basic Hypergeometric Series, vol. 96, 2nd edn. (Cambridge University Press, Cambridge, 2004).
4. A. Tanaka, H. Mori, T. Morita, Superconformal index on $RP^2 \times S^1$ and mirror symmetry. Phys. Rev. **D91**, 105023 (2015). arXiv:1408.3371 [hep-th].

Curriculum Vitae

Data

Name:	Akinori Tanaka
Age:	28
Nationality:	Japanese
Academic degrees:	Mar. 2015, Doctor of Science, Osaka University
Present Affiliation:	Interdisciplinary Theoretical Science Research Group, RIKEN
Address:	Wako 351-0198, JAPAN
E-mail:	akinori.tanaka@riken.jp

Publication List

- M. Honda, N. Iizuka, A. Tanaka, and S. Terashima, "Exact Path Integral for 3D Higher Spin Gravity," arXiv:1511.07546 [hep-th].
- M. Honda, N. Iizuka, A. Tanaka, and S. Terashima, "Exact Path Integral for 3D Quantum Gravity II," Phys. Rev. **D93** (2016) no. 6, 064014, arXiv:1510.02142 [hep-th].
- N. Iizuka, A. Tanaka, and S. Terashima, "Exact Path Integral for 3D Quantum Gravity," Phys. Rev. Lett. **115** no. 16, (2015) 161304, arXiv:1504.05991 [hep-th].
- A. Tanaka, A. Tomiya, and T. Shimotani, "Symmetry breaking caused by large $\mathcal{R}$-charge," J. High Energy Phys. **10** (2014) 136, arXiv:1404.7639 [hep-th].
- A. Tanaka, "Localization on round sphere revisited," J. High Energy Phys. **11** (2013) 103, arXiv:1309.4992 [hep-th].
- A. Tanaka, "Localization and knots," Int. J. Mod. Phys. Conf. Ser. **21** (2013) 195–196.
- A. Tanaka, "Comments on knotted 1/2 BPS Wilson loops," J. High Energy Phys. **07** (2012) 097, arXiv:1204.5975 [hep-th].

A. Tanaka, *Superconformal Index on* $\mathrm{RP}^2 \times \mathrm{S}^1$ *and 3D Mirror Symmetry*,
Springer Theses, DOI 10.1007/978-981-10-1398-0

Zeittracht Medien GmbH
Ferdinand-Jühlke-Straße 7
99095 Erfurt, Deutschland
produktsicherheit@kolibri360.de